A NOMENCLATURE OF COLORS

FOR NATURALISTS,

AND

COMPENDIUM OF USEFUL KNOWLEDGE

FOR ORNITHOLOGISTS.

BY

ROBERT RIDGWAY,

CURATOR, DEPARTMENT OF BIRDS, UNITED STATES NATIONAL MUSEUM

*WITH TEN COLORED PLATES AND SEVEN PLATES
OF OUTLINE ILLUSTRATIONS.*

BOSTON:
LITTLE, BROWN, AND COMPANY.
1886.

Copyright, 1885,
BY ROBERT RIDGWAY.

UNIVERSITY PRESS
JOHN WILSON AND SON, CAMBRIDGE

TO

PROFESSOR SPENCER F. BAIRD,

SECRETARY OF THE SMITHSONIAN INSTITUTION,
AND DIRECTOR OF THE UNITED STATES NATIONAL MUSEUM,

THIS BOOK

IS RESPECTFULLY DEDICATED

BY THE AUTHOR

WASHINGTON, November, 1886

CONTENTS.

	PAGE
INTRODUCTION	9

PART I.

NOMENCLATURE OF COLORS.

PREFACE	15
PRINCIPLES OF COLOR AND GENERAL REMARKS	19
COLORS REQUIRED BY THE ZOÖLOGICAL OR BOTANICAL ARTIST	27
COMPARATIVE VOCABULARY OF COLORS	38
BIBLIOGRAPHY	57

PART II.

ORNITHOLOGISTS' COMPENDIUM.

GLOSSARY OF TECHNICAL TERMS USED IN DESCRIPTIVE ORNITHOLOGY	61
TABLE FOR CONVERTING MILLIMETRES INTO ENGLISH INCHES AND DECIMALS	119
TABLE FOR CONVERTING ENGLISH INCHES AND DECIMALS INTO MILLIMETRES	125

LIST OF PLATES.

PLATES
- I. COMBINATIONS OF PRIMARY AND SECONDARY COLORS
- II GRAYS
- III. BROWNS.
- IV. RED-BROWNS.
- V. BROWN-YELLOWS.
- VI. YELLOWS AND ORANGES.
- VII. REDS.
- VIII. PURPLES.
- IX. BLUES.
- X. GREENS
- XI. FIGURE ILLUSTRATING EXTERNAL ANATOMY OF A BIRD
- XII. FIGURES ILLUSTRATING DETAILS IN EXTERNAL ANATOMY OF A BIRD'S HEAD.
- XIII. FIGURES ILLUSTRATING UNDER SURFACE OF A BIRD'S WING.
- XIV. FIGURES ILLUSTRATING VARIOUS COLOR-MARKINGS.
- XV FIGURES ILLUSTRATING VARIOUS COLOR-MARKINGS.
- XVI. FIGURES ILLUSTRATING VARIOUS EGG-CONTOURS.
- XVII. COMPARATIVE SCALE OF MEASUREMENT STANDARDS.

INTRODUCTION.

THE present volume is intended to supply a want much felt by the author during the course of his ornithological studies, and therefore presumably experienced by other workers in the same field, namely, a nomenclature of colors and a compendious dictionary of technical terms used in descriptive ornithology, together with a series of plates or diagrams illustrating the external anatomy of a bird in relation to the terms employed, and such other things as are more clearly expressed by a picture than by a mere definition. Probably few, if any, naturalists have not on more than one occasion deplored the absence of such an aid to their studies, for it is very difficult, if not impossible, for any one to keep all these things clearly in mind.

Undoubtedly one of the chief desiderata of naturalists, both professional and amateur, is a means of identifying the various shades of colors named in descriptions, and of being able to determine exactly what name to apply to a particular tint which it is desired to designate in an original description. No modern work of this character, it appears, is extant, — the latest publication of the kind which the author has been able to consult being Syme's edition of "Werner's Nomenclature of Colors," published in Edinburgh in 1821,[1] a copy of which the writer has been able to procure through the kind attentions of a correspondent in England. In the selection of plates which accompany the present work, and in forming the definitions to which they refer, the book just cited has been carefully consulted, as have also various others bearing to a greater or less degree upon the same subject. It is found, however, that in Syme's "Nomenclature" the colors have become so modified by time, that in very few cases do they correspond with the tints they were intended to represent. On this account it has not been possible, except in a very few instances, to make the examples given in the present volume agree with those of the book in question, — which is much to be regretted, since as great uniformity as possible is highly desirable in so important a matter. It has occurred to the writer, however, that by careful selection from the fine artists' colors manufactured by the most celebrated makers of the present day (which are believed to be very far superior in purity, as well as much more varied, than those made in Werner's time), some of these may be made the standard by which to fix definitely names for certain tints which otherwise must remain more or less arbitrary. A basis for a fixed nomenclature of colors may thus be obtained, — with this additional advantage, that artists may thereby be furnished a clew to the manufactured colors which are required for

---

[1] Werner's | Nomenclature of Colors, | with additions | arranged so as to render it highly useful to the | Arts and Sciences, | particularly | Zoology, Botany, Chemistry, Mineralogy, and Morbid Anatomy | Annexed to which are | examples selected from well-known objects | in the | Animal, Vegetable, and Mineral Kingdoms | = | By Patrick Syme, | Flower-Painter, Edinburgh, | Painter to the Wernerian and Caledonian | Horticultural Societies. | Second Edition | = Edinburgh ; Printed for William Blackwood, Edinburgh, | and T. Cadell, | Strand, London. | — | 1821. | Small 8vo., pp. 47, 13 pls.

the reproduction of particular tints As having the highest reputation and perhaps the greatest merit, the colors manufactured by Messrs. Winsor & Newton of London, and Fr Schoenfeld & Co. of Dusseldorf, have been chiefly selected as the standards for this work The colors manufactured by these firms embrace so great a variety that it has been found possible to identify with them a large number of those named in descriptions, the mixture of two or more being of course occasionally necessary.

In regard to the external anatomy or "topography" of a bird, a system as little complicated as possible is desirable. The one presented in this work, while substantially the same as that usually adopted, and offering no innovations, is considerably simplified, thereby greatly facilitating the acquirement by the student of a knowledge of this essential adjunct of descriptive ornithology.

It is believed also that the figures representing the typical forms of color-markings, and of egg-contours, and the concordant scale of different standards of measurement, will also be found of great practical utility

The author has in every case endeavored to give the plainest possible definition of a term consistent with accuracy. All expressions having reference solely to internal characters, and which therefore seldom if ever enter into ordinary descriptions of birds, have been excluded, though many anatomical and osteological terms occasionally employed in diagnoses of the higher groups, and others pertaining to the general treatment of the subject, are considered and carefully defined.

Acknowledgments are due from the author to several friends for their generous assistance. Dr LEONHARD STEJNEGER suggested and prepared the comparative scale of standard measurements and the tables for the conversion of English inches to millimetres, and *vice versa*, together with the explanations pertaining thereto. He also aided in the compilation of the comparative color-vocabulary, in which he was substantially assisted by Mr. José C. Zeledon. The outline drawings were executed by Mr. John L. Ridgway from the author's cruder originals.

In the hope that ornithologists in general, and those of the "rising generation" in particular, may find in this volume a convenient and useful book of reference, it is respectfully submitted by

THE AUTHOR.

PART I.

NOMENCLATURE OF COLORS.

PREFACE.

THE want of a nomenclature of colors adapted particularly to the use of naturalists has ever been more or less an obstacle to the study of Nature, and although there have been many works published on the subject of color, they either pertain exclusively to the purely scientific or technical aspects of the case or to the manufacturing industries, or are otherwise unsuited to the special purposes of the zoologist, the botanist, and the mineralogist.

According to a learned authority, who, among others, has been carefully consulted in the preparation of this work, "the names of colors, as usually employed, have so little to do with the scientific or technical aspects of the subject, that we are in reality dealing with the peculiarities of language"[1] This is of course true as considered from the stand-point of pure science; but popular and even technical natural history demands a nomenclature which shall fix a standard for the numerous hues, tints, and shades which are currently adopted, and now form part of the language of descriptive natural history.

It has been the earnest endeavor of the author to attain this object in the present work; and in order to do so he has spared no pains, having for this purpose

[1] VON BEZOLD · Theory of Color, p 99.

procured the finest prepared colors known to modern art, including those of all the best manufacturers, — as Winsor & Newton, George Rowney & Co., and Ackermann, of London, England; Dr Fr Schoenfeld & Co, Dusseldorf; Chenal, Burgeois, Binant, and Lefranc, of Paris; Osborne of Philadelphia, and others. He has, besides, consulted all the authorities accessible to him

In determining the standard for those arbitrary or conventional tints and shades (as chestnut, hair-brown, ash-color, lilac, etc) whose names are taken from some familiar substance or object, which itself varies so much in color that the name without such fixed standard would be practically valueless, care has been taken to select a *characteristic* example

The selection of appropriate names for the colors depicted on the plates has been in some cases a matter of considerable difficulty. With regard to certain ones it may appear that the names adopted are not entirely satisfactory; but, to forestall such criticism, it may be explained that the purpose of these plates is *not to show the color of the particular objects or substances which the names suggest, but to provide for the colors which it has seemed desirable to represent, appropriate, or at least approximately appropriate, names* In other words, certain colors are selected for illustration, for which names must be provided, and when names that are exclusively pertinent or otherwise entirely satisfactory are not at hand, they must be looked up or invented It should also be borne in mind that almost any object or substance varies more or less in color; and that therefore if the "orange," "lemon," or "chestnut" of the plates does not match exactly in color the particular orange, lemon, or chestnut which one may compare it with, it may (or in fact does) correspond with other specimens. It is, in fact, only in the case of those colors which derive their names directly from the pigments which represent them (as Paris green, orange-cadmium, vermilion, ultramarine blue, madder-brown, etc.) that we have absolute pertinence of name to color.

PRINCIPLES OF COLOR

AND

GENERAL REMARKS.

THE popular nomenclature of colors has of late years, especially since the introduction of aniline dyes and pigments, become involved in almost chaotic confusion through the coinage of a multitude of new names, many of them synonymous, and still more of them vague or variable in their meaning. These new names are far too numerous to be of any practical utility, even were each one identifiable with a particular fixed tint. Many of them are invented at the caprice of the dyer or manufacturer of fabrics, and are as capricious in their meaning as in their origin; among them being such fanciful names as "Zulu," "Crushed Strawberry," "Baby Blue," "Woodbine-berry," "Night Green," etc., besides such nonsensical names as "Ashes of Roses" and "Elephant's Breath." An inspection of the sample-books of manufacturers of various fancy goods (such as embroidery silks and crewels) is sufficient to show the absolute want of system or classification which prevails, thus rendering these names peculiarly unavailable for the purposes of science, where absolute fixity of the nomenclature is even more necessary than its simplification.[1]

As is stated on page 23, had we pigments representing the three primary colors in their absolute purity, it would be a simple matter to produce all possible modifications of color by their combination with one another, together with the addition of black or white, when required. Even with the imperfect pigments now available, by far the larger number can be made (see pages 29–32).

According to Von Bezold, the term "hue" is synonymous with color; a "tint" denotes a color or hue modified by admixture of white; while a "shade" implies a color darkened with black. The same author classifies colors as follows: —

I. *Gold, silver, black,* and *white.*
II. *Full colors,* or those of the solar spectrum (that is, blue, green, and red, — or, as some authorities have it, and especially as popularly supposed, blue, yellow, and red).
III. *a. Dark* colors, or those shaded with black. Such may be properly termed "shades" of blue, green, red, etc.
    b. Light colors (diluted or mixed with white) and *pale* colors (which are still further lightened or diluted).
    c. Broken colors, by which is meant "those colors which reach the eye mixed with faint white, that is to say, gray light, but in which the specific character of their hue is still expressed with tolerable decision. If the gray predominates to such an extent that we receive only a very slight sensation of color, we speak of a gray with the addition of the name of a color, such as greenish gray, bluish gray, etc.' (pages 97, 98).

[1] The author is under obligations to the Nonotuck Silk Company, of Florence, Mass., for sample-books of their Corticelli embroidery silks, which at his request were most courteously and gratuitously supplied.

PRINCIPLES OF COLOR.

Without thought of improving upon the above arrangement, the author would nevertheless present the following classification, as perhaps a more convenient one for the purposes of the present work.

I. *Pure colors of the solar spectrum.*
    a. **Primary colors**, or those not produced by mixture.
        1. Red.
        2. Yellow.
        3. Blue.
    b. **Secondary colors**, or those produced by the mixture of two primary colors.
        4. Orange (= red + yellow).
        5. Green (= yellow + blue).
        6. Purple (= blue + red).

II. *Impure colors, or those not found in the solar spectrum.*
    a. **Shades**, which may consist of either primary or secondary colors[1] darkened by black (= complete or absolute shade).
    b. **Tints**, which may consist of either primary or secondary colors lightened by the admixture of white (= absolute degree of light).
    c. **Subdued colors**, which consist of combinations of two or more secondary colors, or of a secondary color with the primary which does not enter into its composition, that is, its complementary color, as green with red, purple with yellow, orange with blue, etc., — the effect being to subdue or neutralize the colors which are thus combined.

---

[1] The principal shades may be classified as follows : —
    a. Shades of primary colors.
        1. Of red (= red + black) = "*maroon.*"
        2. Of yellow (= yellow + black) = "*olive.*"
        3. Of blue (= blue + black) = "*indigo,*" or "*blue-black*"
    b. Shades of secondary colors.
        4. Of orange (= yellow + red + black) = *brown*
        5. Of green (= yellow + blue + black) = *dark green,* "*bottle-green*" "*myrtle-green*"
        6. Of purple (= blue + red + black) = "*plum-purple.*"

It seems scarcely necessary to include the so-called gold and silver colors in the above classification, since they are nearly related, or at least analogous, to yellow and white respectively, the difference consisting chiefly in the existence of a metallic medium or surface.

Observing the colors of the solar spectrum, it is obvious that each secondary color grades insensibly into the two primaries composing it, and that thus results an unbroken transition from one end of the series to the other. The transitions may be shown by the following sequence, the names of the primary colors being given in heavy-faced type (and also preceded by a Roman numeral) and those of the secondary colors in italics.

Spectrum Series.

I. 1. **Red.**
    2. Orange-red.
    3. Reddish orange.
    4. *Orange.*
    5. Yellowish orange.
    6. Orange-yellow.
II. 7. **Yellow.**
    8. Greenish yellow.
    9. Yellowish green.
    10. *Green.*
    11. Bluish green.
    12. Greenish blue.
III. 13. **Blue.**
    14. Purplish blue.
    15. Bluish purple.
    16. *Purple.*
    17. Reddish purple.
    18. Purplish red.

Not only is the transition complete from nos. 1 to 18, but could the names be arranged in the form of a circle, so as to bring the first and last in juxtaposition, these would be found to merge, and thus complete an unbroken ring of graded colors. The three primary colors each enter into the composition of eleven of the eighteen named in the list, as follows. *Red*, in nos 1 to 6 and 14 to 18, inclusive; *yellow*, in nos. 2 to 12; and *blue*, in nos 8 to 18

Returning to the impure colors, or those which do not occur in the solar spectrum, it may be premised that black and white represent, respectively, the absorption and refraction of the sun's rays, the former being in reality a combination of all colors.

It is an axiom of chromatologists that the multitudinous hues, shades, and tints of Nature are simply the results of various combinations of three primary colors, together with the two additional elements of absorption and refraction, making five elements in all, from which it of course follows, as a mathematical deduction, that one hundred and twenty combinations, that is, specific colors (using the latter term in the comprehensive sense) are possible. Additional modifications almost *ad infinitum* are produced by varying circumstances, as different relative proportions of the component elements, effects of contrast, etc

Accepting this theory as correct, it would therefore seem that in order to reproduce from Nature any color, tint, or shade whatsoever, that might be desired, the artist would require only three pigments to represent the primary colors, that is, a red, a yellow,[1] and a blue; together

---

[1] We here speak of yellow as one of the primary colors, for the reason that it is really so to all appearance and intent, so far as the requirements of the artist are concerned. It has, however, we think, been conclusively proven that it is green and not yellow which is the third primary color, in addition to red and blue. Says VON BEZOLD (Theory of Color, p 128).
"Red, yellow, and blue were generally looked upon in former times as the fundamental colors, the results accepted by the mixture of pigments

with black and white, the two latter to represent the elements of absorption and refraction of the sun's rays (or darkness and light respectively). Such would really be the case, and the manipulation of colors therefore a very simple process, were we able to get pigments representing absolutely pure primary colors. Unfortunately, however, the artist's palette does not yet contain even one of them in its requisite purity, neither do the black and white pigments represent satisfactorily the elements of darkness and light. Therefore, it becomes necessary, in order to obtain certain desired results, to make a combination of pigments different from that of the solar spectrum, as, for example, the substitution of yellow for green [1]

We may take hope, however, from the fact that many important discoveries and improvements in the manufacture of artist's colors have been made in the past few years, that the final surmounting of present difficulties may be entirely within the possibilities of chemistry.[2]

---

having been accepted as a basis Later investigations lead to the conclusion that green must be substituted for yellow, and a variety of reasons might be cited, all of which speak unanimously in favor of assuming *red, green,* and a *blue* which borders closely upon violet, to be the fundamental colors" And again, on p. 138 "Yellow was formerly included among the fundamental colors, from purely technical motives This was simply owing to the fact that green can be produced by mixing yellow and blue pigments, while by the mixture of green and red only a very dark yellow, that is to say, a brown, can be obtained"

[1] See preceding foot-note.

[2] Already colors approaching very minutely to the pure hues of the spectrum have been discovered, indeed, they are even manufactured, and to some extent used Unfortunately they are not permanent The aniline reds and purples ("rose-Tyrien," "geranium-red," "solferino," "magenta," "mauve," etc) are of a purity and richness not approached by the madder or cochineal tints, nor by any combination of these with other colors The "rose-carthame," or "saffloiroth," of SCHOENFELD is incomparably purer than the finest vermilions, madder-reds, or carmines, and is perhaps as permanent as the last named, but fades after exposure to

The scope of the present work will not allow an extended dissertation on this subject, the aim being to furnish the student with a convenient means of identifying or determining those colors regarding which he may be more or less uncertain. It is obviously impracticable to illustrate all the numerous hues, shades, and tints which occur in the plumage of birds, but it is believed that the carefully selected assortment depicted on plates II. to X. will answer every reasonable requirement. A great difficulty has been encountered in the arrangement of the colors on the plates, from the circumstance that a *linear series*, which shall express all the relations, gradations, and transitions, is here quite as impossible as in zoological or botanical classifications. Thus, all the purples have more or less of blue and red in their composition; but some of them through the admixture of yellow or gray (black and white) tend more or less toward brown or gray; any other series of compound colors presenting equally perplexing the light; which is unfortunate, since in this color we have almost the exact red of the solar spectrum, and can therefore produce by its combination with the purest yellow (light cadmium) and blue (ultramarine), purer orange and purple tints than can be obtained by the use of any other red. Genuine ultramarine is said to be the most perfect of known pigments, and the same may be said of the lighter cadmium-yellows; so that the great desideratum is a perfect red. Among trustworthy pigments, vermilion, Paris green, and ultramarine are named by Von Bezold (p 136) as those which most nearly represent the primary colors. However, while the two latter are probably as pure as it will be possible to obtain, the first is very far from a perfect red, making neither a pure orange with yellow nor a purple with any blue.

Speaking of this matter, a writer in the "Art Union" (we make the quotation at second hand, from the "Art Interchange," vol xii no 13, p. 148) makes the following observations. "We have a good supply of yellows of every shade, some of them quite durable; we are pretty well furnished with blues, but good reds are very few. The reds of iron [Venetian red, light red, etc] are too dull, the madder preparations are too weak Vermilion is excellent in its place, but there is absolutely no true red of good body and quite durable."

complications In order to make comparison of allied shades and tints more easy, it has been endeavored to place all belonging to a particular class together on one plate, but in not a few cases it has been a difficult matter to decide upon which plate a certain one should be put, the decision being in some instances almost purely arbitrary.

COLORS REQUIRED BY THE ZOOLOGICAL OR BOTANICAL ARTIST.

Notwithstanding so great a variety of colors exist in Nature, especially in the animal and vegetable kingdoms, and that an almost unlimited number of pigments are manufactured for the use of artists, a comparatively very small number is really required. The author has in his collection considerably over three hundred water-colors, each bearing a different name, representing the productions of the best makers (see page 16). Nearly three hundred of them are put aside, however, since very careful experiments have proved that they are superfluous His working palette, selected from the above number, is limited to thirty-six colors, at least one half of which are used for convenience rather than because they are necessary Following is the list, those most essential being distinguished by an asterisk (*)

Black.
*1. Lamp-Black.

Browns.
*2. Bone-Brown, or Bistre.
*3. Roman Brown (Schoenfeld's).
*4. Raw Umber.
*5. Sepia.

Brown-Reds.

*6. Burnt Sienna.
7. Light Red.
8. Neutral Orange.
*9. Indian Red (Schoenfeld's).

Reds.

*10. Madder Carmine (Winsor & Newton's), or Deep Madder Lake (Schoenfeld's).
*11. Scarlet Vermilion.

Orange.

*12 Orange Cadmium (Winsor & Newton's).

Yellows.

13. Middle Cadmium (Schoenfeld's).
*14. Light Cadmium (Schoenfeld's).
15. Ultramarine Yellow (Schoenfeld's) or Lemon Yellow (Winsor & Newton's).
16. Aureolin.

Brown-Yellows.

*17. Yellow Ochre.
*18. Raw Sienna.

Greens.

19 Terre Verte (Winsor & Newton's.)
*20. Dark Zinnober Green (Schoenfeld's).
21. Light Zinnober Green (Schoenfeld's).
*22. Viridian (Winsor & Newton's).
23. Green Oxide Chromium (Schoenfeld's).
24 Paris Green.
*25. Emerald Green.

Blues.

26. Intense Blue (Winsor & Newton's).
*27. Antwerp Blue.

*28. Violet Ultramarine (Schoenfeld's).
*29. French Ultramarine.
 30. Italian Ultramarine (Rowney's).
*31. Cobalt Blue.

Gray

 32. Payne's Gray.

Purples

 33. Violet Krapplack (Schoenfeld's).
 34 Blue Krapplack (Schoenfeld's)
*35. Purple Madder (Winsor & Newton's).

White.

*36. Chinese White (Schoenfeld's "gouache-farben").

Colors (including some of the above-named) which are sometimes considered necessary, may be readily produced by combination of others, as follows —

Burnt Umber,	by mixture of sepia or bistre with burnt sienna.				
Raw Umber,	,,	,, .,	,,	,,	raw sienna.
Vandyke Brown,	,,	,, ,,	,,	,,	burnt sienna.
Ultramarine Blue,	,,	,, ,,	French blue and cobalt.		
Small Blue,	,,	,, ,,	,, violet ultramarine		
Paris Green,	,,	,, emerald green and light cadmium.			
Violet Madder-lake,	,,	,, French blue and madder-carmine.			
Terre-verte Green,	,,	,, viridian with black and white.			

As may be inferred from the circumstance that the author has, in the above list, occasionally indicated his preference for the particular "make," colors of the same name vary in tone or quality according to the manufacturer Thus, the olive-greens of Winsor & Newton and Schoenfeld respectively are conspicuously different, being equally useful, however. It is quite likely that the wares of each manufacturer may vary to a greater or less extent, but the writer's actual experience is indicated by the preferences noted above. Certain it is that he has uniformly found Schoenfeld's Permanent Chinese White ("gouache-farben"), put up in small wide-mouthed glass bottles, superior in working and keeping qualities to Winsor & Newton's preparation of the same name. It should further be explained that for convenience English names are given in the above list to Schoenfeld's colors, and that the names by which they are labelled may be found in the comparative vernacular synonymy (giving English, French, and German names for each pigment) on pages 38–55.

Were the cochineal and aniline colors permanent, the above list would have to be increased by the addition of carmine, rosalack (light), mauve (aniline-violet), rose Tyrien, and dark aniline-green;[1] since, with possibly the exception of the first, it is impossible to imitate them by combinations of other colors, so great are their purity and intensity. Rose-carthame (safflorroth, or safflower-red), a vegetable color, is incomparably purer than any variety of vermilion or carmine; in fact, it is the only red which will, combined with yellow and blue respectively, produce both a pure orange and purple. It has the reputation of being evanescent, however, and therefore, like the aniline and cochineal colors, should not be used where permanence of color is an object, unless in cases where the pictures thus colored are to be only occasionally, and for short periods at a time, exposed to the light.

---

[1] The nearest approach to mauve that can be attained by mixture of permanent colors is that produced by combination of permanent blue, or Italian ultramarine, with madder-carmine or madder-lake. Carmine may be quite closely imitated by mixture of madder-carmine and scarlet-vermilion. Rose-carthame, rosalack, rose Tyrien, and dark aniline-green are absolutely imitable; so, for that matter, is mauve.

A very large number of pigments which are in general use, but which are really superfluous, can be exactly imitated by mixtures of those named in the foregoing list, for example: —

The *cochineal reds* (crimson-lake, carmine, scarlet-lake, etc.), by mixture of madder-carmine (or deep madder-lake) and scarlet-vermilion, in proper proportion.

Red-lead, Saturn-red, and *orange-chrome,* by combination of scarlet-vermilion and orange-cadmium; the colors thus produced being decidedly superior in working qualities to the pigments they are intended to replace, while they are at least equal in brilliancy.

Purple-lake may be imitated by mixture of madder-carmine and lamp-black.

Dragon's-blood red, by light vermilion and lamp-black.

Mars violet, by ultramarine blue (or Italian ultra) and light vermilion.

Burnt madder-lake, by madder-carmine and permanent blue.

Purple (Schoenfeld's), by madder-carmine and Antwerp blue.

Madder-violet (Chenal's), by Antwerp blue and rose-madder.

Rubens's madder, by madder-carmine and burnt sienna.

Brown madder, by madder-carmine, burnt sienna, and sepia.

Burnt carmine, by madder-carmine and lamp-black.

Violet carmine, by madder-carmine, lamp-black, and Antwerp blue.

Dahlia carmine, by madder-carmine and lamp-black.

Indigo, by Italian ultra or permanent blue and lamp-black.

Middle cadmium, by orange-cadmium and pale cadmium.

Olive-green (Schoenfeld's), by Italian ultra, lamp-black, pale cadmium, and sepia.

Olive-green (Winsor & Newton's), by Antwerp blue, aureolin, lamp-black, and sepia.

Dark aniline-blue and *violet-ultramarine,* by mixture of French blue and madder-carmine.

Azure-blue, by combinations of Italian ultra or permanent blue and Chinese white

Turquoise-blue, *celestial blue* (*colinblau* of Schoenfeld), *rock-blue* (*bergblau* of Schoenfeld), *cerulean blue*, and *blue oxide*, by mixture of Antwerp blue and Chinese white.

Blue-verditer and *green-blue oxide*, by *Antwerp blue, light cadmium*, and *Chinese white*.

Green-lake, by viridian and lamp-black

The foregoing are only a few examples, and the list might be increased almost indefinitely; but these will suffice

Regarding the selection of colors by an artist, an authority in the "Art Union" says:—

"Among the pigments prepared by the modern colormen, many of the most attractive are utterly untrustworthy. We will say nothing of the brilliant aniline colors which are so showy and yet will scarcely last a day, but we will select three colors which are in constant use, and which it seems almost impossible to get along without These are chrome-yellow, carmine-red, and Prussian blue. Samples of these hung in a strong light will within a year completely lose their color, turning green and black"

With very few exceptions. all of the colors depicted on the plates of this work can be produced from a palette of five pigments, — black, white, red, yellow, and blue. For convenience, however, the primary colors should be combined into secondaries (orange, green, and purple), while a gray and a brown should be added, the first produced by mixture of black and white, the second by combination of red and green, — making all together ten elements, as follows:—

1. Black (lamp-black).
2. White (Chinese white).
3. Red (madder-carmine or deep madder-lake + scarlet-vermilion).[1]
4. Orange (scarlet-vermilion + Schoenfeld's light cadmium).
5. Yellow (Schoenfeld's light cadmium).
6. Green (Schoenfeld's light cadmium + Italian ultramarine).
7. Blue (Italian ultramarine).
8. Purple (Italian ultramarine + madder-carmine).
9. Gray (lamp-black + Chinese white).
10. Brown (red + green).

With these ten elements ninety binary combinations may be made, resulting in as many more or less distinct colors, the number of which may be increased almost indefinitely by varying the relative proportion of the component parts. The following is a list of these combinations, together with the names of the resultant colors: —

a Modifications of Black.

11. Black + white = slate.
12. „ + red = seal-brown.
13. „ + orange = clove-brown.
14. „ + yellow = dark olive-green.
15. „ + green = greenish black.
16. „ + blue = bluish black; indigo.
17. „ + purple = purplish black.

[1] In compounding a purple, the madder-red should be used, and not vermilion, while in preparing an orange, the latter should be used and not the former. These two reds are necessary, for the reason that they form the nearest approach to a pure red among pigments that can be relied on for permanence. Neither of them, however, will by itself serve all the purposes for which a pure red is necessary, since a pure orange cannot be made with the madder-reds, nor a purple with vermilion. Rose-carthame or saffloroth (safflower-red) is of the requisite purity, but is said to lack permanence.

18. Black + gray = slate-black.
19. ,, + brown = brownish black.

b. Modifications of White.

20. White + black = gray.
21. ,, + red = pink.
22. ,, + orange = yellowish salmon-color.
23. ,, + yellow = primrose-yellow.
24. ,, + green = pea-green
25. ,, + blue = pale blue.
26. ,, + purple = lilac.
27. ,, + gray = pale gray.
28. ,, + brown = Isabella-color.

c. Modifications of Red.

29. Red + black = burnt carmine.
30. ,, + white = reddish pink.
31. ,, + orange = orange-red.
32. ,, + yellow = orange-red.
33. ,, + green = brownish red; brick red.
34. ,, + blue = reddish purple.
35. ,, + purple = purplish red
36. ,, + gray = grayish purple.
37. ,, + brown = brownish red, brick red.

d. Modifications of Orange.

38. Orange + black = russet-olive.
39. ,, + white = creamy orange.
40. ,, + red = reddish orange, intense orange.
41. ,, + yellow = yellowish orange.
42. ,, + green = yellowish ochraceous.
43. ,, + blue = brownish ochraceous.
44. ,, + purple = tawny ochraceous.
45. ,, + gray = ochraceous-buff.
46. ,, + brown = orpiment-orange; brownish orange.

e. Modifications of Yellow.

47. Yellow + black   = olive-green.
48.   ,,   + white   = canary-yellow.
49.   ,,   + red   = orange.
50.   ,,   + orange = orange-yellow.
51.   ,,   + green   = citron-yellow.
52.   ,,   + blue   = yellowish green.
53.   ,,   + purple = wax-yellow.
54.   ,,   + gray   = olive-yellow.
55.   ,,   + brown   = saffron-yellow.

f. Modifications of Green.

56. Green + black   = bottle-green.
57.   ,,   + white   = malachite-green.
58.   ,,   + red   = sage-green.
59.   ,,   + orange = olive-green.
60.   ,,   + yellow = yellowish green.
61.   ,,   + blue   = bluish green.
62.   ,,   + purple = dark sage-green.
63.   ,,   + gray   = grayish green.
64.   ,,   + brown   = olive.

g. Modifications of Blue.

65. Blue + black   = marine-blue.
66.   ,,   + white   = cobalt-blue; azure-blue.
67.   ,,   + red   = violet.
68.   ,,   + orange = dark sage-green.
69.   ,,   + yellow = bluish green, sea-green.
70.   ,,   + green   = greenish blue.
71.   ,,   + purple = purplish blue; hyacinth-blue.
72.   ,,   + gray   = grayish blue
73.   ,,   + brown   = indigo.

h. Modifications of Purple.

74. Purple + black   = auricula-purple.
75.   ,,   + white = lilac-purple.

76. Purple + red   = reddish purple; magenta.
77.   „   + orange = brownish purple.
78.   „   + yellow = brownish purple.
79.   „   + green = grayish purple; livid purple.
80.   „   + blue   = violet.
81.   „   + gray   = grayish purple.
82.   „   + brown = brownish purple; Indian purple.

i. Modifications of Gray.

83. Gray + black = slate-color.
84. „   + white = pale gray.
85. „   + red   = vinaceous-gray.
86. „   + orange = Isabella-drab.
87. „   + yellow = yellowish gray; olive-gray.
88. „   + green = greenish gray.
89. „   + blue   = bluish gray.
90. „   + purple = purplish gray.
91. „   + brown = brownish gray; drab-gray.

j. Modifications of Brown.

92. Brown + black = dark brown; bistre.
93.   „   + white = wood-brown.
94.   „   + red   = reddish brown.
95.   „   + orange = russet.
96.   „   + yellow = yellowish brown.
97.   „   + green = olive-brown.
98.   „   + blue   = dark olive.
99.   „   + purple = purplish brown; Vandyke brown.
100. „   + gray   = grayish brown; drab.

NOTE. — It should be remembered that each of the above represents a combination distinct from all the others. For example, "red + black" and "black + red" imply very different relative proportions of the two colors, the former being *black* modified by admixture of a small quantity of red, the latter being *red* modified by the addition of a little black.

The following colors have after careful experiment been found to be unsafe, as being liable to fade or change in time, or produce chemical reaction when combined with others, their absolute rejection by the artist is therefore advised· The *chrome-yellows*, for which the cadmiums should be substituted, the *chrome-greens*, which may be exactly imitated by mixture of Antwerp blue and light cadmium; all the *cochineal colors* (carmine, crimson-lake, purple-lake, and scarlet-lake); all the *aniline colors*, including the pigments known as geranium-red (geranium-lack of Schoenfield), rosalack, solferino, magenta, mauve, etc.; *rose-carthame* (safflorroth), *yellow lake*, *Italian pink*, *brown-pink*, *pure scarlet* (which is completely and very rapidly evanescent), *guano red* and *Prussian blue*. Gamboge is also of doubtful permanence, but there is no other equally pure transparent yellow known. The list of unreliable colors is a very large one; therefore, instead of giving it in full, the author will merely caution the reader against the use of any of those mentioned above, and at the same time assure him that by adopting the "palette" recommended on pages 27–29 he will be able to reproduce almost any color that he may have occasion to imitate

COMPARATIVE

GIVING EQUIVALENT NAMES IN ENGLISH, LATIN, GERMAN,

English.	Latin.	German.
Amethyst.	Amethystinus.	Amethyst.
*Antwerp Blue.[2]		Antwerpner blau.
*Apple Green.		Apfelgrün.
*Aquamarine Blue.	Aquamarinus.	
*Ash-color (see Cinereous).	Cinereus; cineraceus.	Aschfarbig.
Ashy.	Cinerascens.	Aschfarbig.
*Aster Purple.		
Aureolin.		
*Auricula Purple.		Aurikel-purpur.
*Azure Blue (see Sky Blue).	Azureus; cœlicolor.	Azur blau.
*Bay.	Badius.	
*Bice Green.		
*Bistre.		Bister.
*Berlin Blue.		Berliner blau.
*Beryl Green.	Berylinus.	Beryl-grün.
*Black.	Ater; niger.	Schwarz.
Blackish.	Nigrescens.	Schwärzlich.
Blackish Blue.	Atro-cœruleus; atro-cyaneus.	Schwarzblau.
Blackish Brown.	Atro-brunneus.	Schwarzbraun.
Blackish Crimson.	Atro-carmesinus.	Schwärzlich carmesin.
Blackish Green.	Atro-viridis; nigro-viridis.	Schwarzgrün.
Blackish Olive.	Atro-olivaceus.	Swärzlich olivenfarbig.
Blackish Purple.	Atro-purpureus.	Schwärzlich purpurfarbig.
Blackish Slate.	Atro-schistaceus.	Schwärzlich schieferfarbig.
Blackish Violet.	Atro-violaceus.	Schwärzlich violet.
Blood Red.	Sanguineus; sanguineo-ruber.	Blutroth.
Blue.	Cyaneus; cœruleus.	Blau.
Bluish.	Cyanescens; cœrulescens.	Bläulich.
Bluish Black.	Cyanater.	Blauschwarz.
Bluish Gray.	Cyaneo-canus.	Blaugrau.
Bluish Green.	Cyaneo-viridis.	Blaugrün.
Bluish Slate.	Cyano-schistaceus.	Bläulich schieferfarbig.
Bluish Violet.	Cyano-violaceus.	Blau violet.
Bluish White.	Cyano-albidus.	Bläulich weiss.
*Bottle Green.		Flaschengrün.
*Brick Red (see Tile Red).	Lateritius; testaceus; rutilus.	Ziegelroth.
*Broccoli Brown.		Broccolibraun.
Bronze.	Æneus.	Bronze.

[1] In the preparation of this vocabulary I have received very valuable assistance from my friends
[2] Colors distinguished by a * are represented on plates I.-X.

VOCABULARY OF COLORS,

FRENCH, SPANISH, ITALIAN, NORWEGIAN, AND DANISH.[1]

French.	Spanish.	Italian.	Norwegian and Danish.
Améthyste.	Amatista.	Ametista.	Ametyst.
Bleu d'Anvers.	Azul de Ambéres.		Antwerpen-blaa.
Vert de pomme.	Verde manzana.	Verde di pomo.	Æble-grön.
Cendré.	Cinéreo.	Cinereo.	Askefarvet.
Cendré.	Ceniciento.	Cenerino.	Aske-.
Pourpre d'auricula.	Púrpura de aurícula.	Porporino di auricula.	Aurikel-purpur.
Bleu d'azur; bleu céleste.	Azul celeste.	Azzurro.	Asur-blaa.
Bai.	Bayo.	Baio.	
Bistre.			Bister.
Bleu de Berlin.	Azul de Berlin.	Azzuro.	Berliner-blaa.
Vert béril.	Verde berilo.	Verde berillino.	Beryl-grön.
Noir.	Negro.	Nero; negro.	Sort.
Noirâtre.	Negruzco.	Nerastro; nericcio.	Sortagtig.
Bleu noirâtre.	Azul negruzco.	Azzurro nerastro.	Sorte-blaa.
Brun noirâtre.	Moreno negruzco.	Bruno nerastro.	Sorte-brun.
Cramoisi noirâtre.	Carmesí negruzco.	Chermesino nerastro.	Sorte-karmesin.
Vert noirâtre.	Verde negruzco.	Verde nerastro.	Sorte-grön.
Couleur d'olive noirâtre.	Aceitunado negruzco.	Olivaceo nerastro.	Sortagtigt olivenfarvet.
Pourpre noirâtre.	Púrpura negruzco.	Porpora nerastro.	Sortagtigt purpurfarvet.
Ardoisé noirâtre.	Pizarra negruzco.	Ardesiaco nerastro.	Sortagtigt skiferfarvet.
Violet noirâtre.	Violeta negruzco.	Violetto nerastro.	Sorte-violet.
Rouge de sang.	Rojo sangre.	Rosso sanguineo.	Blod-röd.
Bleu.	Azul.	Azzurro.	Blaa.
Bleuâtre.	Azuloso.	Azzurrognolo.	Blaa-agtig.
Noir bleuâtre.	Negro azuloso.	Nero azzurrognolo.	Blaa-sort.
Gris bleuâtre.	Gris azuloso.	Grigio azzurrognolo. [gnolo.	Blaa-graa.
Vert bleuâtre.	Verde azuloso.	Verde azzurrognolo.	Blaa-grön.
Ardoisé bleuâtre.	Pizarra azuloso.	Ardesiaco azzurrognolo.	Blaagtig skiferfarvet.
Violet bleuâtre.	Violeta azuloso.	Violetto azzurrognolo.	Blaa-violet.
Blanc bleuâtre.	Blanco azuloso.	Bianco azzurrognolo.	Blaa-hvid.
Vert de bouteille.	Verde botella.	Verde-bottiglia.	Flaske-grön.
Rouge de brique.	Rojo ladrillo.	Rosso di mattone.	Mursteus-röd.
Brun di broccoli.	Moreno de bróculi.	Bruno di broccoli.	Broccoli-brun.
Broncé.	Bronce.	Bronzo.	Bronse.

Mr. LEONHARD STEJNEGER, of Bergen, Norway, and Sr. DON JOSÉ C. ZELEDON, of San José, Costa Rica.

COMPARATIVE VOCABULARY

ENGLISH.	LATIN.	GERMAN.
Bronzy.	Æneus.	Bronzirt.
Bronze Green.	Æneo-viridis.	Bronzegrün.
Bronze Purple.	Æneo-purpureus.	Bronze purpur.
Brown.	Brunneus.	Braun.
Brownish.	Brunnescens.	Bräunlich.
Brownish Black.	Brunneo-niger.	Braunschwarz.
Brownish Buff.	Brunneo-luteus.	Bräunlich chamois.
Brownish Gray.	Brunneo-canus.	Braun grau.
Brownish Green.	Brunneo-viridis.	Braungrün.
Brownish Ochraceous.	Brunneo-ochraceus.	Bräunlich ockerfarbig.
Brownish Olive.	Brunneo-olivaceus.	Bräunlich olivenfarbig.
Brownish Orange.	Brunneo-aurantius.	Bräunlich orange.
Brownish Pink.	Brunneo-carneus.	Bräunlich nelkenfarbig.
Brownish Purple.	Brunneo-purpureus.	Bräunlich purpurfarbig.
Brownish Red.	Brunneo-ruber.	Braunroth.
Brownish Slate.	Brunneo-schistaceus.	Bräunlich schieferfarbig.
Brownish Vinaceous.	Brunneo-vinaceus.	Bräunlich weinfarben.
Brownish White.	Brunneo-albidus.	Bräunlich weiss.
Brownish Yellow.	Brunneo-flavus.	Braun-gelb.
*Buff.	Luteus; luteolus.	Chamois.
*Buff-Yellow.	Luteo-flavus.	Chamois-gelb.
*Buff-Pink.	Luteo-caryophyllaceus.	
Buffy Brown.	Luteo-brunneus.	Chamois-braun.
Buffy Gray.	Luteo-griseus.	Chamois-grau.
Buffy Ochraceous.	Luteo-ochraceus.	Chamois ockerfarbig.
Buffy White.	Luteo-albidus.	Chamois-weiss.
*Burnt Carmine.		Gebrannter carmin.
*Burnt Sienna.		Gebrannte terra di Sienna.
*Burnt Umber.		Gebrannte umbra.
*Cadmium Orange.	Cadmiumino-aurantius.	Orange cadmium.
*Cadmium Yellow.	Cadmiumino-flavus.	Cadmium-gelb.
*Canary Yellow.		
*Campanula Blue.		Campanula-blau.
*Carmine.	Carmineus; coccineus.	Carmin.
*Cerulean Blue.	Cœruleus; cœlicolor.	Coelin blau.
*Chestnut.	Castaneus.	Kastanienfarbig.
Chestnut Brown.	Castaneo-brunneus.	Kastanienbraun.
Chestnut-Rufous.	Castaneo-rufus.	Kastanienroth.
*China Blue.		
*Chinese Orange.		Chinisch orange.
*Chocolate.	Chocolatinus.	Chocoladenfarbig.
*Chrome Yellow.		

COMPARATIVE VOCABULARY.

OF COLORS. — *Continued*.

FRENCH.	SPANISH.	ITALIAN.	NORWEGIAN AND DANISH.
Broncé.	Bronceado.	Bronzato.	Bronse-; bronseret.
Vert broncé.	Verde bronceado.	Verde bronzato.	Bronse-grön.
Pourpre broncé.	Púrpura bronceado.	Porpora bronzata.	Bronse-purpur.
Brun.	Moreno.	Bruno.	Brun.
Brunâtre.	Morenuzco.	Brunastro.	Brunagtig.
Brun noir; noir brunâtre.	Negro morenuzco.	Nero brunastro.	Brun-sort.
Chamois brunâtre.	Ante morenuzco.	Camoscio brunastro.	Brunagtig læderfarvet; brunagtig blak.
Gris brunâtre; bruncendré.	Gris morenuzco.	Grigio brunastro.	Brun-graa.
Vert brunâtre.	Verde morenuzco.	Verde brunastro.	Brun grön.
Ocre brune.	Ocre morenuzco.	Ocraceo brunastro.	Brunagtig okerfarvet.
Olive brunâtre.	Aceitunado morenuzco.	Olivaceo brunastro.	Brunagtig olivenfarvet.
Orange brunâtre.	Naranjado morenuzco	Aranciato brunastro.	Brunagtig orangefarvet.
Rose clair brunâtre.	Encarnado morenuzco	Roseo-chiaro brunastro.	Brunagtig nellikeröd.
Pourpre brunâtre.	Púrpura morenuzco.	Porporino brunastro.	Brunagtig purpurfarvet.
Brun-rouge.	Rojo morenuzco.	Rosso brunastro.	Brun-röd.
Ardoisé brunâtre.	Pizarreño morenuzco.	Ardesiaco brunastro.	Brunagtig skieferfarvet.
Vineux brunâtre.	Moreno vinoso.	Vinato brunastro.	Brunagtig vinfarvet.
Blanc brunâtre.	Blanco morenuzco.	Bianco brunastro.	Brunagtig hvid.
Jaune brunâtre.	Amarillo morenuzco.	Giallo brunastro.	Brun-gul.
Chamois.	Ante.	Colore camoscio.	Chamois.
Jaune chamois.	Amarillo de ante.	Giallo-camoscio.	Chamois-gul.
Brun chamois.	Moreno de ante.	Bruno-camoscio.	Chamois.
Gris chamois.	Gris de ante.	Grigio-camoscio.	Chamois-graa.
Ocre chamois.	Ocre de ante.	Ocraceo-camoscio.	Chamois.
Blanc chamois.	Blanco de ante.	Bianco-camoscio.	Chamois.
Carmin brulé.	Carmin quemado.	Carmino bruciato.	Brændt karmin.
Terre de Sienna brulée.	Tierra de Siena quemada.	Terra di Sienna bruciata.	Brændt sienna.
Terre d'Ombre brulée.	Tierra de sombra quemada.	Terra di Ombria bruciata.	Brændt umber.
Cadmium orange.	Cadmio naranjado.	Aranciato di cadmio.	Kadmium-orange.
Jaune de cadmium.	Amarillo de cadmio.	Giallo di cadmio.	Kadmium-gul.
Bleu de campanule.	Azul de campánula.	Azzurro di campanella.	Blaaklokke-blaa.
Carmin.	Carmin.	Carmino.	Karmin.
Bleu céleste.	Azul celeste.	Ceruleo; celeste; celestino.	Himmel-blaa.
Châtain.	Castaño.	Castagno; castaguino.	Kastanje-; kastanjefarvet.
Brun de châtain.	Moreno castaño.	Bruno castagnino.	Kastanje-brun.
Roux de châtain.	Rojizo castaño.	Rosso castagnino.	Kastanje-röd.
Orange chinois.	Naranjado chino	Aranciato di Cina.	Kinesisk orange.
Couleur de chocolat.	Chocolate.	Bruno-cioccolata.	Chokoladefarvet.

A NOMENCLATURE OF COLORS.

COMPARATIVE VOCABULARY

ENGLISH.	LATIN.	GERMAN.
*Cinereous (see Ash-color).	Cinereus.	Aschenfarbig; asch-grau.
*Chromium Green.		
*Cinnamon.	Cinnamomeus; cinnamominus.	Zimmtfarbig.
Cinnamon Brown.	Cinnamomeo-brunneus.	Zimmtbraun.
*Cinnamon-Rufous.	Cinnamomeo-rufus.	Zimmtroth.
*Citron Yellow.		
*Claret Brown.	Vinaceo-brunneus.	Weinbraun.
*Clay-color.	Luteus; lutosus.	Lehmfarbig.
*Clove Brown.		
*Cobalt Blue.	Cobaltinus.	Kobaltblau.
Coppery.	Cupreus; cuprescens.	Kupferfarbig.
Coppery Bronze.	Cupreo-æneus.	Kupferbronze.
Coppery Red.	Cupreo-ruber.	Kupferroth.
*Coral Red.	Corallinus; corallino-ruber.	Koralroth.
*Cream-color.		Rahmfarbig.
*Creamy Buff.		
Creamy White.		Rahmweiss.
Creamy Yellow.		Rahmgelb.
*Crimson.	Carmesinus.	Carmesin.
*Dahlia Purple.		Dahlia purpur.
*Deep Chrome.		
*Drab.		
Drab-Brown.		
*Drab-Gray.		
*Dragon's-blood Red.		Drachenblut.
*Emerald Green.	Smaragdinus.	Smaragdgrün.
*Fawn-color.	Cervinus; cervineus.	Hirschfarbig.
*Écru-Drab.		
*Ferruginous (see Rusty).	Ferrugineus.	Rostfarbig.
Ferruginous Brown.	Ferrugineo-brunneus.	Rostbraun.
Ferruginous Chestnut.	Ferrugineo-castaneus.	Rostfarbig kastanienbraun.
Ferruginous Rufous.	Ferrugineo-rufus.	Rostroth.
*Flax-flower Blue.		
*Flesh-color.	Carneus; incarnatus.	Fleischfarbig.
*Flame-scarlet.	Flammeus; igneus.	
*French Blue.		Französisch blau.
*French Gray.		Französisch grau.
*French Green.		
Fuliginous (see Sooty).	Fuliginosus.	Russfarbig.
Fuliginous Black.	Fuliginoso-niger.	Russ-schwarz.
Fuliginous Brown.	Fuliginoso-brunneus.	Russ-braun.
Fuliginous Gray.	Fuliginoso-griseus; fuliginoso-caneus; fuliginoso-cinereus.	Russ-grau.
Fuliginous Olive.	Fuliginoso-olivaceus.	Russfarbig oliven.
Fulvous (see Tawny).	Fulvus.	

COMPARATIVE VOCABULARY. 43

OF COLORS. — *Continued.*

FRENCH.	SPANISH.	ITALIAN.	NORWEGIAN AND DANISH.
Cendré.	Ceniciento.	Cinereo.	Aske-graa.
Couleur cannelle.	Canela.	Canella.	Kanel-; Kanelfarvet.
Brun cannelle.	Moreno canela.	Bruno canellino.	Kanel-röd.
Roux cannelle.	Rojizo canela.	Rosso canellino.	Kanel-brun.
Brun vineux.	Moreno vinoso.	Bruno vinato.	Vin-brun.
Terre clay.	Arcilloso.	Colore d'argilla; lutoso.	Ler-; lerfarvet; blak.
Brun de clou de girofle.	Pardo de clavo.	Bruno de garofano.	Krydder-nellik-brun.
Bleu de cobalt.	Azul de cobalto.	Azzurro cobalto.	Kobalt-blaa.
Cuivre.	Cobrizo.	Rameico.	Kobber-; kobberfarvet.
Bronce de cuivre.	Bronce cobrizo.	Bronzo rameico.	Kobber-bronze.
Rouge de cuivre.	Rojo cobrizo.	Rosso rameico.	Kobber-röd.
Rouge de corail; coralin.	Rojo coral.	Corallino; rosso corallino.	Koral-röd.
Couleur de la crême.	Color de crema.	Colore di crema.	Flöde-; flödefarvet.
	Crema de ante.	Camoscio di crema.	
Blanc nuancé de couleur de la crême.	Blanco crema.	Bianco di crema.	Flöde-hvid.
Jaune de la crême.	Amarillo crema.	Giallo di crema.	Flöde-gul.
Cramoisi.	Carmesí.	Chermesino.	Karmesin.
Pourpre dahlia.	Púrpura de dalia.	Porpora di dalia.	Dahlia purpur.
Couleur de drap.	Color de paño.		Drap.
Brun de drap.	Moreno de paño.		Drap-brun.
Gris de drap.	Gris de paño.		Drap-graa.
Sang de dragon.	Sangre de drago.	Sangue di drago.	Drageblod-röd.
Vert émeraude.	Verde esmeralda.	Verde smeraldo.	Smaragd-grön.
Brun cervine.	Cervino.	Cervino; lionato.	Hjorte-; hjortefarvet.
Ferrugineux.	Ferruginoso.	Ferruginoso.	Rust-; rustfarvet.
Brun ferrugineux.	Moreno ferruginoso.	Bruno ferruginoso.	Rust-brun.
Châtain ferrugineux.	Castaño ferruginoso.	Castagno ferruginoso.	Rustfarvet kastanjebrun.
Roux ferrugineux.	Rojizo ferruginoso.	Rosso ferruginoso.	Rust-röd.
Couleur de chair.	Encarnado.	Color carneo.	Kjödfarvet.
Bleu français.	Azul francés.	Azzurro francese. Grigio francese.	Fransk-blaa.
Gris français.	Gris francés.		Fransk-graa.
Fuligineux.	Fuliginoso.	Fuligginoso.	Sod-; sodfarvet.
Noir fuligineux.	Negro fuliginoso.	Nero fuligginoso.	Sod-sort.
Brun fuligineux.	Moreno fuliginoso.	Bruno fuligginoso.	Sod-brun.
Gris fuligineux.	Gris fuliginoso.	Grigio fuligginoso.	Sod-graa.
Olive fuligineux.	Aceitunado fuliginoso.	Olivaceo fuligginoso.	Sodfarvet oliven.
Fauve.	Leonado.	Fulvo.	

A NOMENCLATURE OF COLORS.

COMPARATIVE VOCABULARY

ENGLISH.	LATIN.	GERMAN.
Fuscous.	Fuscus.	Dunkelfarb.
*Gallstone Yellow.		
*Gamboge Yellow.		Gummi-guttæ.
Garnet Red.	Granatinus.	Granat-roth.
*Geranium Red.		Geraniumroth.
*Geranium Pink.		Geraniumrosa.
Glaucous.	Glaucus.	
*Glaucous Blue.	Glauco-cœruleus.	
Glaucous Gray.	Glauco-griseus; glauco-canus; glauco-cinereus.	
*Glaucous Green.	Glauco-viridis.	
Glaucous White.	Glauco-albidus.	
Golden.	Aureus.	Golden.
Golden Bronze.	Aureo-æneus.	Goldbronze.
Golden Green.	Aureo-viridis.	Goldgrün.
Golden Yellow.	Aureo-flavus.	Goldgelb.
*Grass Green.	Prasinus.	Grassgrün.
Gray.	Canus; griseus; cæsius; leucophæus.	Grau.
Grayish.	Canescens.	Graulich.
Grayish Black.	Cano-niger; griseo-ater.	Grauschwarz.
Grayish Blue.	Cano-cyaneus; cyanescens; cœrulescens.	Graublau.
Grayish Brown.	Cano-brunneus.	Graubraun.
Grayish Buff.	Griseo-lutosus; griseo-luteus.	Graulich fahl.
Grayish Green.	Cano-viridis.	Graugrün.
Grayish Olive.	Cano-olivaceus.	Graulich olivenfarbig.
Grayish Purple.	Cano-purpureus.	Graulich purpurfarben.
Grayish Violet.	Cano-violaceus.	Grauviolet.
Grayish White.	Cano-albidus; griseo-albidus.	Graulich weiss.
Grayish Yellow.	Griseo-flavus.	Graulich gelb.
Green.	Viridis.	Grün.
Greenish.	Virescens; viridescens.	Grünlich.
Greenish Black.	Viridi-ater.	Grünschwarz.
Greenish Blue.	Viridi-cyaneus.	Grünblau.
Greenish Brown.	Viridi-brunneus.	Grünlich braun.
Greenish Buff.	Viridi-luteus.	Grünlich chamois.
Greenish Gray.	Viridi-canus.	Grüngrau.
Greenish Olive.	Viridi-olivaceus.	Grünlich olivenfarbig.
Greenish Slate.	Viridi-schistaceus.	Grünlich schieferfarbig.
Greenish White.	Viridi-albus.	Grünlich weiss.
Greenish Yellow.	Viridi-flavus.	Grüngelb.
*Hair Brown.		Haarbraun.
*Hazel.	Coryllinus; avellinus.	Hazelbraun; nussbraun.
*Heliotrope Purple.		
Hoary.	Pruinosus; canescens; albescens.	Frostgrau.
Horn-color.	Corneus.	Hornfarbig.

OF COLORS. — *Continued.*

FRENCH.	SPANISH.	ITALIAN.	NORWEGIAN AND DANISH.
Sombre.	Oscuro.	Fosco.	Mörkfarvet.
Gomme gutte.	Goma guta.		Gummi-gut.
Rouge de grenat.	Rojo de granate.	Rosso di granato.	Granat-rød.
Rouge de geranium.	Rojo de geranio.	Rosso di geranio.	Geranium-rød.
Rouge claire de geranium.	Rosado de geranio.	Rosso chiaro di geranio.	Geranium-rosa.
	Glauco.	Glauco.	
	Azul glauco.	Azzuro glauco.	Kaal-blaa.
	Gris glauco.	Grigio glauco.	
	Verde glauco.	Verde glauco.	
	Blanco glauco.	Bianco glauco.	
Couleur d'or.	Aureo.	Dorato.	Guld-; gylden.
Or faux.	Bronce dorado.	Bronzo dorato.	Guld-bronse.
Vert d'or.	Verde dorado.	Verde dorato.	Guld-grøn.
Jaune d'or.	Amarillo dorado.	Giallo dorato.	Guld-gul.
Vert d'herbe; vert végétal.	Verde hierba.	Verde di erba.	Græs-grøn.
Gris.	Gris.	Grigio; cano.	Graa.
Grisâtre.	Grisoso.	Grigiastro.	Graa-agtig.
Noir grisâtre; gris-noir.	Negro grisoso.	Nero grigiastro.	Graa-sort.
Bleu grisâtre.	Azul grisoso.	Turquino-grigiastro.	Graa-blaa.
Brun grisâtre.	Moreno grisoso.	Bruno-grigiastro.	Graa-brun.
Chamois grisâtre.	Ante grisoso.	Cano-lutoso.	Graa-blak.
Vert grisâtre.	Verde grisoso.	Verde grigiastro.	Graa-grøn.
Olive grisâtre.	Aceitunado grisoso.	Olivacco grigiastro.	Graa-agtig olivenfarvet.
Pourpre grisâtre.	Púrpura grisoso.	Porporino grigiastro.	Graa-agtig purpurfarvet.
Violet grisâtre.	Violeta grisoso.	Violetto grigiastro.	Graa-violet.
Blanc grisâtre.	Blanco grisoso.	Bianco grigiastro.	Graa-hvid.
Jaune grisâtre.	Amarillo grisoso.	Giallo grigiastro.	Graa-gul.
Vert.	Verde.	Verde.	Grøn.
Verdâtre.	Verdoso.	Verdastro; verdiccio.	Grønagtig.
Noir verdâtre.	Negro verdoso.	Nero verdastro.	Grøn-sort.
Bleu verdâtre.	Azul verdoso.	Verd-azzuro.	Grøn-blaa.
Brun verdâtre.	Moreno verdoso.	Bruno verdastro.	Grønlig brun.
Chamois verdâtre.	Ante verdoso.	Camoscio verdastro.	Grønagtig chamois.
Gris verdâtre.	Gris verdoso.	Grigio-verdastro.	Grøn-graa.
Olive verdâtre.	Aceitunado verdoso.	Olivacco verdastro.	Grønagtig olivenfarvet.
Ardoisé verdâtre.	Pizarreño verdoso.	Ardesia verdastro.	Grønagtig skieferfarvet.
Blanc verdâtre.	Blanco verdoso.	Bianco verdastro.	Grøn-hvid.
Jaune verdâtre.	Amarillo verdoso.	Giallo verdastro.	Grøn-gul.
Brun de cheveux.	Moreno de pelo.	Bruno di capello.	Haar-brun.
Brun-noisette.	Moreno de avellana.	Color nocciola.	Hassel-brun; nødbrun.
Grison.	Canoso.		Frost-graa; graaskinlet.
Couleur de corne.	Color de cuerno.	Color di corno.	Horn-; hornfarvet.

A NOMENCLATURE OF COLORS.

COMPARATIVE VOCABULARY

ENGLISH.	LATIN.	GERMAN.
*Hyacinth Blue.	Hyacinthinus.	Hyacinthpurpur.
*Indian Purple.		Indisch purpur.
Indian Red.		Indischroth.
*Indian Yellow.		Indischgelb.
*Indigo Blue.	Indigoticus.	Indigo blau.
*Isabella-color.	Isabellinus.	Isabellafarbig.
King's Yellow.		King's gelb.
*Lake Red.		Lack.
*Lavender.	Lavendulaceus.	Lavendelfarbig.
*Lavender Gray.	Lavendulaceo-canus.	Lavendelgrau.
Lavender Pink.	Lavendulaceo-carneus.	Lavendelroth.
Lavender Purple.	Lavendulaceo-purpureus.	Lavendelpurpur.
Lead-color (see Plumbeous).	Plumbeus.	Bleifarbig.
*Lemon Yellow.	Citrinus; citreus.	Citrongelb.
*Lilac.	Lilacinus; lilaceus.	Lila.
*Lilac Gray.	Lilacino-canus.	Lila grau.
Lilac Pink.	Lilacino-carneus.	Lila-fleischfarben.
Lilac Purple.	Lilacino-purpureus.	Lilapurpur.
Livid.	Lividus.	Todtenblau.
*Liver Brown.	Hepaticus.	Leberbraun.
*Madder Brown.		Madderbraun; brauner krapp.
Madder Purple.		Madderpurpur; purpur krapp.
*Magenta.		Anilinrosa.
*Maize Yellow.		
*Malachite Green.		Malachitgrün.
*Marine Blue.		Marinblau.
*Mars Brown.		
*Maroon.	Atro-purpureus; atro-cocci-[neus.	
*Maroon-Purple.		
*Mauve.	Malvinus.	Hellviolet.
*Mouse Gray.	Murinus.	Mäuse-grau.
*Mummy Brown.		
*Myrtle Green.		Myrthengrün.
*Naples Yellow.		Neapel gelb.
Neutral tint.		Neutral tinte.
*Ochraceous.	Ochraceus.	Ocker; ockerfarbig.
Ochraceous Brown.	Ochraceo-brunneus. [luteus.	Ockerbraun.
*Ochraceous Buff.	Ochraceo-luteosus; ochraceo-	Ockerfahl.
*Ochraceous Rufous.	Ochraceo-rufus.	Ockerroth.
Ochraceous White.	Ochraceo-albus.	Ockerweiss.
*Ochraceous Yellow.	Ochraceo-flavus.	Ockergelb.
*Oil Green.	Oleagineo-viridis.	Oelgrün.
Oil Yellow.	Oleagineo-flavus.	Oelgelb.
*Olive.	Olivaceus; olivinus.	Olivenfarbig.
Olive-Brown.	Olivaceo-brunneus.	Olivenbraun.
*Olive-Buff.	Olivaceo-luteus.	Olivenfahl.
Olive-Drab.		
*Olive-Gray.	Olivaceo-canus.	Olivengrau.
*Olive-Green.	Olivaceo-viridis.	Olivengrün.

COMPARATIVE VOCABULARY.   47

OF COLORS.— *Continued.*

FRENCH.	SPANISH.	ITALIAN.	NORWEGIAN AND DANISH.
Pourpre d'hyacinthe.	Púrpura de jacinto.	Porpora giacinto.	Hyasint-purpur.
Pourpre indienne.	Púrpura de India.	Porpora indica.	Indisk purpur.
Rouge indienne.	Rojo de India.	Rosso indico.	Indisk röd.
Jaune indienne.	Amarillo de India.	Giallo indico.	Indisk gul.
Indigo.	Indigo.	Indigo.	Indigo-blaa.
Isabelle.	Amarillo de Ysabel.	Colore isabella; isabellino.	Isabel-; isabelfarvet.
Jaune de King.	Amarillo de King.	Giallo di King.	King's gult.
Laque.	Laca.	Lacca.	Lack-röd.
Couleur de lavande.	Alhucema.	Lavanda.	Lavendel-; lavendelfarvet.
Gris de lavande.	Gris alhucema.	Grigio lavanda.	Lavendel-graa.
Rouge de lavande.	Encarnado alhucema.	Rossico lavanda.	Lavendel-röd.
Pourpre de lavande.	Púrpura alhucema.	Porpora lavanda.	Lavendel-purpur.
Couleur de plomb.	Aplomado.	Plumbeo.	Bly-; blytarvet.
Jaune citron.	Amarillo limon.	Citrino.	Citron-; citron-gul.
Lilas.	Lila.	Lilacino.	Lila.
Gris lilas.	Gris lila.	Lilacino-grigio.	Lila-graa.
Rose-claire lilas.	Encarnado lila.	Lilacino-carneo.	Lila-kjödfarvet.
Pourpre lilas.	Púrpura lila.	Lilacino-porporino.	Lila-purpur.
Livide.	Lívido.	Livido.	Blaagusten.
Brun hépatique.	Moreno higado.	Epatico.	Lever-brun.
Brun de Madder; Garance brune.	Moreno de rubia.	Bruno di robbia.	Krap-brun.
Pourpre de Madder; Garance pourpre.	Púrpura de rubia.	Porporino di robbin.	Krap-purpur.
Magenta.	Magenta.	Magenta.	Magenta-röd.
Vert malachite.	Verde malaquita.	Verde di malachito	Malakit-grön.
Bleu marine.	Azul marino.	Azzurro marino.	Marine-blaa.
Marron.	Moreno carmesí.	Marrone.	Maron.
Mauve.	Malva.	Malvino.	Malve-violet.
Couleur de souris.	Gris de raton.	Colore di sorice.	Muse-grau.
Vert de myrte.	Verde mirto.	Verde mirtino.	Myrte-grön.
Jaune de Naples.	Amarillo de Nápoles.	Giallo di Napoli.	Neapel-gul.
Teinte neutre.			
Ocre.	Ocráceo.	Ocraceo.	Oker-; okerfarvet.
Brun ochracée.	Moreno ocráceo.	Bruno ocraceo.	Oker-brun.
Chamois ochracée.	Gamuza ocráceo.	Camoscio ocraceo.	Oker-blak.
Rouge ochracée.	Rojizo ocráceo.	Rosso ocraceo.	Oker-röd.
Blanc ochracée.	Blanco ocráceo.	Bianco ocraceo.	Oker-hvid.
Jaune ochracée.	Amarillo ocráceo.	Giallo ocraceo.	Oker-gul.
Vert d'huile.	Verde aceite.	Verde di olio.	Olje-grön.
Jaune d'huile.	Amarillo de aceite.	Giallo di olio.	Olje-gul.
Olivâtre.	Aceitunado.	Olivaceo; olivastro.	Oliven-; olivenfarvet.
Brun-olivâtre.	Moreno aceitunado.	Bruno olivastro.	Oliven-brun.
Chamois olivâtre.	Ante aceitunado.	Camoscio olivastro.	Oliven-blak.
Couleur de drap olivâtre.	Color de paño aceitunado.		
Gris olivâtre.	Gris aceitunado.	Grigio olivastro.	Oliven-graa.
Vert olive.	Verde aceitunado.	Verde olivastro.	Oliven-grön.

A NOMENCLATURE OF COLORS.

COMPARATIVE VOCABULARY

English.	Latin.	German.
*Olive-Yellow.	Olivaceo-flavus.	Olivengelb.
*Orange.	Aurantius.	Orangefarbig; pomeranzfarbig.
Orange-Brown.	Aurantio-brunneus.	Orangebraun. / Pomeranzbraun.
*Orange Chrome.		
*Orange-Buff.	Aurantio-luteus.	Orangefahl. / Pomeranzfahl.
*Orange-Ochraceous.		
Orange-Red.	Aurantio-ruber.	Orangeroth. / Pomeranzroth.
*Orange-Rufous.	Aurantio-rufus.	
*Orange-Vermilion.		
Orange-Yellow.	Aurantio-flavus.	Orangegelb. / Pomeranzengelb.
*Orpiment Orange.		Auripigment-orange.
Pale Blue.		Hellblau.
*Pansy Purple.		Pensé-purpur.
*Paris Blue.		Pariser blau.
*Paris Green.		Pariser grün.
*Parrot Green.	Viridissimus.	Papageigrün.
*Pea Green.		Erbsengrün.
*Peach-blossom Pink.		Pfirsichblumenroth.
*Pearl Blue.		Perlblau.
*Pearl Gray.		Perlgrau.
*Phlox Purple.		
Pink.	Caryophyllaceus; pallide roseus.	Nelkenroth.
Pinkish.		Nelkenröthlich.
Pinkish Brown.		Nelkenröthlich braun.
*Pinkish Buff.		Nelkenröthlich fahl.
Pinkish Flesh-color.		
Pinkish Lilac.		Nelkenröthlich lila.
Pinkish Orange.		Nelkenröthlich orangefarbig.
Pinkish Red.		
Pinkish White.		Nelkenröthlich weiss.
*Pinkish Vinaceous.		
*Plum Purple.		Pflaumenpurpur.
*Pomegranate Purple.	Puniceus; phœniceus.	
*Poppy Red.		Mohrroth; Ponceau.
*Primrose Yellow.	Primulaceo-flavus.	
*Prout's Brown.		
*Prune Purple.		Zwetschenpurpur.
*Plumbeous.	Plumbeus.	
Prussian Blue.		Preussisch blau.
Prussian Green.		Preussisch grün.
Purple.	Purpureus.	Purpur.
Purplish.	Purpurascens.	Purpur-.
Purplish Black.	Purpureo-niger.	Purpurschwarz.

COMPARATIVE VOCABULARY.

OF COLORS. — *Continued.*

FRENCH.	SPANISH.	ITALIAN.	NORWEGIAN AND DANISH.
Jaune olivâtre.	Amarillo aceitunado.	Giallo olivastro.	Oliven-gul.
Orange.	Naranjado.	Aranciato.	Orange; brand-gul.
Brun orangé.	Moreno naranjado.	Bruno aranciato.	Orange-brun.
Chamois orangé.	Ante naranjado.	Camoscio aranciato.	Orange-blak.
Rouge orangé.	Rojo naranja.	Rosso aranciato.	Orange-röd.
Roux orangé.	Rojizo naranjado.	Rosso aranciato.	Orange-rödbrun.
Jaune orangé.	Amarillo naranja.	Giallo aranciato.	Orange-gul.
Orange d'orpiment.	Naranjado oropimento	Aranciato di orpimento.	Auripigment orange.
Bleu clair; bleu pâle.	Azul claro.	Azzurro-chiaro.	Lyseblaa.
Pourpre de la pensée.	Púrpura de pensamiento.		Pensé-purpur.
Bleu de Paris.	Azul de Paris.	Azzurro di Paris.	Pariser-blaa.
Vert de Paris.	Verde de Paris.	Verde di Paris.	Pariser-grön.
Vert perruche; vert paroquet.	Verde papagallo.	Verde di pappagallo.	Papegöje-grön.
Vert de pois.	Verde guisante.	Verde di pisello.	Ærte-grön.
Rose fleur de pêcher.	Flor de durazno.	Colore fiore di pesca.	Fersken-blomst-röd.
Bleu perle.	Azul de perla.	Azzurro-perla.	Perle-blaa.
Gris perle.	Gris de perla.	Grigio-perla.	Perle-graa.
Rose-claire.	Rosado claro.	Roseo-chiaro.	Nellik-röd.
			Nellik-rödlig.
Pourpre de la prune.	Púrpura de ciruela.	Porpora di prugna.	Blomme-purpur.
Ponceau.	Punzó.	Ponso.	Valmue-röd; ponceau.
Jaune primevère.	Amarillo de prímula.	Giallo di fiore di primavere.	Primel-gul.
Pourpre de pruneau.	Púrpura de ciruela pasa.	Porpora di sussina.	Sveske-purpur.
Bleu prussien.	Azul de Prusia.	Azzurro di Prussia.	Pröjsisk-blaat.
Vert prussien.	Verde de Prusia.		Pröjsisk-grönt.
Pourpre.	Púrpura.	Porpora; porporino.	Purpur; purpur-farvet.
Tirant sur le pourpre.	Purpúreo.	Porporeggiante.	Purpur-.
Noir pourpré.	Negro purpúreo.	Nero porporeggiante; porporino-nero.	Purpur-sort.

COMPARATIVE VOCABULARY

ENGLISH.	LATIN.	GERMAN.
Purplish Blue.	Purpureo-cyaneus.	Purpurblau.
Purplish Brown.	Purpureo-brunneus.	Purpurbraun.
Purplish Buff.	Purpureo-luteus.	Purpurfahl.
Purplish Gray.	Purpureo-canus.	Purpurgrau.
Purplish Red.	Purpureo-ruber; puniceus; phœniceus.	Purpurroth.
Purplish Rufous.	Purpureo-rufus.	
Purplish Slate.	Purpureo-schistaceus.	Purpur-schieferfarben.
Purplish White.	Purpureo-albidus.	Purpur.
Raisin Purple.		Rosinenpurpur.
*Raw Sienna.		Terra di Sienna.
*Raw Umber.	Umbrinus.	Umbra; bergbraun.
Red.	Ruber.	Roth.
Reddish.	Rubellus; rubescens.	Röthlich.
Reddish Black.	Rubro-niger.	Röthlich schwarz.
Reddish Brown.	Rubro-brunneus.	Rothbraun.
Reddish Buff.	Rubro-luteus.	Röthlich fahl.
Reddish Gray.	Rubro-canus.	Röthlich grau.
Reddish Orange.	Rubro-aurantius.	Röthlich orangefarben.
Reddish Pink.	Rubro-caryophyllaceus.	
Reddish Purple.	Rubro-purpureus; puniceus; phœniceus.	Röthlich purpurfarben.
*Red Lead (see Saturn Red).	Miniatus; flammeus; igneus.	Mennige; saturnroth.
*Rose-color (see Rose Red).	Roseus; rosaceus; rosaceo-ruber.	Rosenfarbig.
*Rose Pink.	Rosaceo-incarnatus; caryophyllaceus; pallide-roseus.	Blassrosa.
*Rose Purple.	Rosaceo-purpureus.	Rosa purpur.
*Rose Red.	Rosaceo-rubrum.	Rosenroth.
*Royal Purple.	Ianthinus.	Königspurpur.
Ruby Red.	Rubineus.	Rubinroth.
Rufescent.	Rufescens.	
*Rufous.	Rufus.	
Rufous-Brown.	Rufo-brunneus.	
Rufous-Buff.	Rufo-luteus.	
Rufous Orange.	Rufo-aurantius.	
*Russet.	Russus.	
Russet Olive.	Russo-olivaceus.	
Russet Drab.		
Rusty (see Ferruginous).	Ferrugineus.	Rostfarben.
Safflower Red (see Geranium Red).		Safflorroth.
*Saffron Yellow.	Croceus; crocco-flavus.	Safrangelb.
*Saturn Red (see Red Lead).	Miniatus.	
*Sage Green.		Salbeigrün.
*Salmon-Buff.		
*Salmon-color.	Salmonaceus.	Lachsfarben.
*Sap Green (see Grass Green)		Saftgrün.
*Scarlet.	Scarlatinus.	Scharlach.
*Scarlet Vermilion.	Scarlatino-cinnabarinus.	Scharlach-zinnoberroth.
*Sea Green.	Thalassinus.	Seegrün.
*Seal Brown.		Pelzrobbenbraun.

COMPARATIVE VOCABULARY.

OF COLORS. — *Continued.*

FRENCH.	SPANISH.	ITALIAN.	NORWEGIAN AND DANISH.
Bleu pourpré.	Azul purpúreo.		Purpur-blaa.
Brun pourpré.	Moreno purpúreo.		Purpur-brun.
Chamois pourpré.	Ante purpúreo.		Purpur-blak.
Gris pourpré.	Gris purpúreo.		Purpur-graa.
Rouge pourpré.	Rojo purpúreo.		Purpur-röd.
Roux pourpré.	Rojizo purpúreo.		Purpur-rödbrun.
Ardoisé pourpré.	Pizarreño purpúreo.		Purpur-skiferfarvet.
Blanc pourpré.	Blanco purpúreo.		
Pourpre de raisin sec.	Púrpura de uva pasa.	Purpura di uva passa.	Rosin-purpur.
Terre de Sienna.	Tierra de Siena.	Terra di Sienna.	Raa sienna.
Terre d'Ombre.	Tierra de sombra.	Terra di ombria.	Raa umber.
Rouge.	Rojo.	Rosso.	Röd.
Rougeâtre.	Rojizo.	Rossastro; rossiccio; rossigno.	Rödlig; rödagtig; röd-.
Noir rougeâtre.	Negro rojizo.		Rödlig-sort.
Brun rougeâtre.	Moreno rojizo.	Bruno rossastro.	Röd-brun.
Chamois rougeâtre.	Ante rojizo.		
Gris rougeâtre.	Gris rojizo.		
Orange rougeâtre.	Naranjado rojizo.		Rödlig orange.
Pourpre rougeâtre.	Púrpura rojizo.		Rödlig purpur.
Rouge de Saturne; minium.	Rojo de Saturno.	Minio.	Mönje-röd.
Rose.	Rosado.	Colore roseo; rosaceo; rosato.	Rosenfarvet; rosen-.
Rose clair.	Encarnado rosado.		Nellike-röd.
Pourpre rosé.	Púrpura rosado.		Rosen-purpur.
Rouge rosé.	Rojo rosado.	Rosso-rosaceo.	Rosen-röd.
Pourpre royal.	Púrpura real.	Porpora reale.	Konge-purpur.
Rouge rubis.	Rojo rubí.	Rosso-robino; robinozzo. Rufescente.	Rubin-röd.
Roux.	Rojizo.	Rossiccio.	Fuks-röd.
Brun roux.	Moreno rojizo.		Fuks-brun.
Chamois roux.	Ante rojizo.		
Orange roux.	Naranja rojizo.		Fuks-orange.
Roux.	Bermejo.	Rossiccio.	
Roux de rouille.	Herrumbrado.	Rugginoso.	Rustfarvet.
Rose carthame.	Rojo de cártamo.	Rosso cartamo.	
Jaune de safran.	Amarillo de azafrán.	Croceo.	Safran-gul.
Vert de sauge.	Verde salvia.	Verde di salvia.	Salvie-grön.
Couleur de saumon.	Color de salmon.	Rosso-salmone.	Laksfarvet.
Vert végétal; vert de vessie.	Verde vegetal.	Verde di succhio.	Saft-grön.
Écarlate.	Escarlata.	Scarlatto.	Skarlagen.
Vermillon écarlate.	Vermellon escarlata.	Vermiglio-scarlatto	Skarlagen-sinober.
Vert de mer.	Verde mar.	Verde marino.	Sjö-grön; hav-grön.
Brun de phoque.	Moreno de foca.	Bruno di foca.	Pelskobbe-brun.

A NOMENCLATURE OF COLORS.

Comparative Vocabulary

ENGLISH.	LATIN.	GERMAN.
*Sepia.	Sepia.	Sepia.
*Sevres Blue.		
Silvery.	Argentatus; argenteus; argentaceus.	Silberfarben.
Silvery Gray.	Argentaceo-canus.	Silbergrau.
Silvery White.	Argentaceo-albus.	Silberweiss.
*Sky Blue (see Azure).	Azureus; cœruleus; cœlicolor; cœlestinus.	Himmelblau; azur blau.
*Slate-color.	Schistaceus; ardosiaceus.	Schieferfarbe.
Slate Black.	Schistaceo-niger.	Schieferschwarz.
Slate Blue.	Schistaceo-cyaneus.	Schieferblau.
Slate Brown.	Schistaceo-brunneus.	Schieferbraun.
*Slate Gray.	Schistaceo-canus.	Schiefergrau.
Slate Green.	Schistaceo-viridis.	Schiefer-grün.
Slate Purple.	Schistaceo-purpureus.	Schieferpurpur.
*Smalt Blue.		Smalte.
Smoky.	Fumosus.	Rauchfarben.
*Smoke Gray.	Fumoso-canus.	Rauchgrau.
Snowy.	Nivosus; niveus; nivalis.	Schneeweiss.
Snuff Brown.		Schnupftabakbraun.
*Solferino Purple.		Anilinrosa.
Sooty (see Fuliginous).	Fuliginosus.	Russfarben.
*Straw Yellow.	Stramineus.	Strohgelb.
Steel Blue.	Chalybæus.	Stahlblau.
*Sulphur Yellow.	Sulphureus.	Schwefelgelb.
*Tawny (see Fulvous).	Fulvus; mustelinus.	Lederfarbig.
*Tawny ochraceous.	Fulvo-ochraceus.	
Testaceous (see Brick Red).	Testaceus; lateritius.	Ziegelroth.
*Tawny Olive.	Fulvo-olivaceus.	
Tile Red (see Brick Red).	Lateritius; testaceus.	Ziegelroth.
*Terre-verte Green.		
*Turquoise Blue.	Turcoso-cyaneus; turcosus.	Turkisblau.
Topaz.		Topas.
*Ultramarine Blue.	Ultramarinus; lazulinus.	Ultramarinblau.
*Vandyke Brown.		Van Dyck-braun.
Venetian Red.		Venetia roth.
*Verdigris Green.		Verdigris.
*Verditer Blue.		Verditer blau.
*Vermilion.	Cinnabarinus.	Zinnober; hochroth.
*Vinaceous.	Vinaceus.	Weinfarbig.
Vinaceous Brown.	Vinaceo-brunneus.	Weinbraun.
*Vinaceous Buff.	Vinaceo-luteus.	Weinfahl.
*Vinaceous Cinnamon.	Vinaceo-cinnamomeus.	Weinröthlich zimmt.
Vinaceous Gray.	Vinaceo-canus.	Weinröthlich grau.
*Vinaceous Pink.	Vinaceo-incarnatus.	Weinröthlich rosa.
Vinaceous Purple.	Vinaceo-purpureus.	Weinpurpur.
*Vinaceous Rufous.	Vinaceo-rufescens.	
Vinaceous White.	Vinaceo-albidus.	Weinröthlich weiss.
*Violet.	Violaceus; ianthinus.	Violet.
Violet-Black.	Violaceo-niger.	Violet-Schwarz.
Violet-Blue.	Violaceo-cyaneus.	Violet-Blau.
Violet-Brown.	Violaceo-brunneus.	Violet-Braun.
Violet-Gray.	Violaceo-canus.	Violet-Grau.

OF COLORS. — *Continued.*

FRENCH.	SPANISH.	ITALIAN.	NORWEGIAN AND DANISH.
Sepia naturelle.	Sepia.	Seppia.	Sepia.
Couleur d'argent.	Plateado.	Argenteo; argentino; d' argento.	Sölv-.
Gris d'argent.	Gris plateado.	Grigio argentino.	Sölv-graa.
Blanc d'argent.	Blanco plateado.		Solv-hvid.
Bleu céleste.	Azul celeste.	Celeste; celestino. [di ardesia.	Himmel-blaa.
Ardoisé.	Pizarreño.	Lavagnato; colore	Skifer-; skiferfarvet.
Noir ardoisé.	Negro pizarreño.	Nero-lavagna.	Skifer-sort.
Bleu ardoisé.	Azul pizarreño.		Skifer-blaa.
Brun ardoisé.	Moreno pizarreño.		Skifer-brun.
Gris ardoisé.	Gris pizarreño.	Grigio-lavagna.	Skifer-graa.
Vert ardoisé.	Verde pizarreño.		Skifer grön.
Pourpre ardoisé.	Púrpura pizarreño.		Skifer-purpur.
Smalt.	Azul de esmalte.	Azzurro di smalto.	Smalt.
Fumeux.	Ahumado.	Fumicoso; fumido.	Rög-; rögfarvet.
Gris fumeux.	Gris ahumado.	Grigio-fumido.	Rög-graa.
Blanc de neige.	Blanco de nieve.	Nevoso.	Sne-; sne-hvid.
Brun de râpé.	Moreno rapé.	Bruno di rapè.	Snus-brun.
Solferino.	Solferino.	Solferino.	Solferino-röd.
Couleur de sonié.	Fuliginoso.	Fuligginoso.	Sod-; sodfarvet.
Jaune de la paille.	Amarillo paja.	Giallo-pagliato.	Straa-gul.
Bleu d'acier.	Azul de acero.	Azzurro acciaio.	Staal-blaa.
Jaune de soufre.	Amarillo de azufre.	Giallo-zolfino.	Svovl-gul.
Basané.	Prieto.	Fulvo.	Læderfarvet.
Testacé.	Testaceo.	Testaceo.	Tegl-röd.
Rouge de brique.	Rojo ladrillo.	Rosso di tegola.	Tegl-röd.
Bleu turquoise.	Azul turquesa.	Turquino.	Turkis-blaa.
Topaze.	Topacio.	Topazio.	Topas.
Bleu outremer.	Azul ultramarino.	Azzurro oltremare.	Ultramarin blaa.
Brun Vandyk.	Moreno de Van Dyck.	Bruno di Van Dyck.	Van-Dyck-brun.
Rouge venetien.	Rojo de Venecia.	Rosso di Venezia.	Venetia röd.
Vert de gris.	Cardenillo.	Verderame.	Verdigris.
Bleu verditer.	Azul verdoso.	Azzurro verdaccio.	Verditer-blaa.
Vermillon; cinabre.	Vermellon.	Vermiglio.	Vermiljon; sinober; cinnober.
Couleur de vin; vineux.	Vinaceo.	Vinato.	Vin-; vinfarvet; vin-röd.
Brun vineux.	Moreno vinaceo.	Bruno vinato.	Vin-brun.
Chamois vineux.	Ante vinaceo.		Vin-blak.
Couleur de cannelle vinacée.			Vinfarvet kanel-brun.
Gris vineux.	Gris vinaceo.		Vin-graa.
Rose clair vineux.	Encarnado vinaceo.		Vin-rod.
Pourpre vineux.	Púrpura vinaceo.		Vin-purpur.
Blanc vineux.	Blanco vinaceo.		Vinfarvet hvid.
Violet.	Violeta.	Violaceo; violetto.	Viol-; violet.
Noir violet.	Negro violeta.	Nero-violacco.	Violet-sort.
Bleu violet.	Azul violeta.	Azzurro violaceo.	Violet-blaa.
Brun violet.	Moreno violeta.		Violet-brun.
Gris violet.	Gris violeta.		Violet-graa.

COMPARATIVE VOCABULARY

ENGLISH.	LATIN.	GERMAN.
Violet Purple.	Violaceo-purpureus.	Violet-Purpur.
Violet Ultramarine.	Violaceo-ultramarinus.	
Veronese Green.		
*Viridian Green.		Viridian grün.
*Wax Yellow.		
*Walnut Brown.		Wallnuss-braun.
Warbler Green (see Olive Green).	(Olivaceo-viridis.)	Laubsänger-grün (Olivengrün).
White.	Albus.	Weiss.
Whitish.	Albescens; albidus.	Weisslich.
Whitish Buff.	Albo-luteus.	Weisslich fahl.
Wine Brown.	Vinaceo-brunneus.	Wein braun.
*Wine Purple.	Vinaceo-purpureus.	Wein purpur.
Wine Red.	Vinaceo-rubrum.	Weinroth.
*Wood Brown.		
Yellow.	Flavus.	Gelb.
Yellowish.	Flavescens; flavicans; flavidus.	Gelblich.
Yellowish Brown.	Flavo-brunneus.	Gelbbraun.
Yellowish Buff.	Flavo-luteus.	Gelbfahl.
Yellowish Drab.		
Yellowish Green.	Flavo-viridis.	Gelbgrün.
*Yellowish Ochraceous.	Flavo-ochraceus.	
Yellowish Olive.	Flavo-olivaceus.	Gelblich olivenfarbig.
Yellowish Orange.	Flavo-aurantius.	Gelblich orangefarbig.
Yellowish White.	Flavo-albus.	Gelblich weiss.
Zinc Yellow.		Zinkgelb.

COMPARATIVE VOCABULARY.

OF COLORS. — *Concluded.*

FRENCH.	SPANISH.	ITALIAN.	NORWEGIAN AND DANISH.
Pourpre violet.	Purpureo violáceo.		Violet-purpur.
Vert Paul Véronèse.	Verde de Veronese.	Verde Paulo Veronese.	Veronese-grön.
Vert émeraude.	Verde viridian.	Verde viridian.	Viridian-grön.
Brun de noix. (Vert olivâtre.)	Moreno de nogal.	Bruno di noce.	Valnöd-brun. Lövsanger-grön.
Blanc.	Blanco.	Bianco; albo.	Hvid.
Blanchâtre.	Blanquecino.	Biancastro; bianchiccio; albescenti; albiccio.	Hvidagtig.
Chamois blanchâtre.	Ante blanquecino.		Hvid-blak.
Brun vineux.	Moreno vinaceo.	Bruno-vinato.	Vin-brun.
Pourpre vineux.	Púrpura vinaceo.		Vin-purpur.
Rouge vineux.	Rojo vinaceo.		Vin-röd.
Jaune.	Amarillo.	Giallo.	Gul.
Jaunâtre.	Amarillento.	Giallastro; gialliccio.	Gulagtig.
Brun jaunâtre.	Moreno amarillento.	Bruno giallastro.	Gul-brun.
Chamois jaunâtre.	Gamuza amarillento.		Gul-blak.
Couleur de drap jaunâtre.			
Vert jaunâtre.	Verde amarillento.		Gul-grön. [vet-
Olive jaunâtre.	Oliva amarillento.		Gulagtig oliven-far-
Orange jaunâtre.	Naranja amarillento.		Gulagtig orange.
Blanc jaunâtre.	Blanco amarillento.		Gul-hvid.
Jaune de zinc.	Amarillo de zinc.	Giallo di zinco.	Zink-gul.

BIBLIOGRAPHY.

The more important works specially consulted in this connection are the following, mentioned nearly in order of their relative importance: —

Bezold, Dr. William von. — The | Theory of Color | in its relation to | Art and Art-Industry | By | Dr. William von Bezold | Professor of Physics at the Royal Polytechnic School of Munich, and member | of the Royal Bavarian Academy of Sciences. | Translated from the German | by | S R Koehler. | With an Introduction and Notes | by | Edward C. Pickering | Thayer Professor of Physics at the Massachusetts Institute of Technology. | Authorized American edition. | Revised and enlarged by the Author. | Illustrated by chromo-lithographic Plates and Wood-cuts. | Vignette. | Boston . | L. Prang and Company. | 1876 | 8vo. pp. iii–xxxii, 1–274, pls. xi. [Without nomenclature of colors.]

Rood, Ogden N. — The International Scientific Series. | —— | Students' | Text-Book of Color; | or, | Modern Chromatics, | with | Applications to Art and Industry. | By | Ogden N. Rood, | Professor of Physics in Columbia College, New York. | With 130 original illustrations | New York . | D. Appleton and Company, | 1, 3, and 5 Bond Street. | 1881. Small 8vo, pages 329 ; 1 colored plate.

Herrick, H. W. — Water Color Painting : | Description of Materials | with | directions for their use in elementary Practice | — | Sketching from Nature in Water Color. | — | By | H W. Herrick. | — | "Artists' Edition." Containing hand-washed examples of one hundred and twenty colors on

Water-Color Paper. | — | New York : | F. W. Devoe & Co., | Corner Fulton and William Streets. | — | 1882. | 12mo, pages i–vii, 9–128, pls. I, II, A—I.

Hay, D. R. — A | Nomenclature of Colours | applicable to the | Arts and Natural Sciences | to | Manufactures | and other Purposes of General | Utility | Second Edition improved | William Blackwood and Sons | Edinburgh and London | MDCCCXLVI | [8vo., cloth, containing 40 full-page colored plates illustrating 228 colors, hues, tints, and shades, but these not named according to the requirements of Natural History.]

Syme, Patrick — Werner's | Nomenclature of Colors, | with additions, | arranged so as to render it highly useful | to the | Arts and Sciences, | particularly | Zoology, Botany, Chemistry, Mineralogy, | and Morbid Anatomy. | Annexed to which are | examples selected from well-known objects | in the Animal, Vegetable, and Mineral Kingdoms. | = | By | Patrick Syme, | Flower-Painter, Edinburgh, | Painter to the Wernerian and Caledonian | Horticultural Societies | Second Edition | = | Edinburgh · | Printed for William Blackwood, Edinburgh ; | and T. Cadell, Strand, London. | — | 1821. 8vo, pages 47.

Martel, Charles. — The | Principles of Colouring | in Painting. | By Charles Martel. | "To imitate the model faithfully, we must copy it differently from what we see it." Chevreul | Twelfth Edition. | Ars probat ^{Crest} artificem | London : | Winsor & Newton, 38, Rathbone Place, | Manufacturing Artists' Colourmen, by Special Appointment to Her Majesty | and Their Royal Highnesses the Prince and Princess of Wales.

Radde, Otto. — Radde's | Internationale Farbenskala | 42 Gammen mit circa 900 Tonen. [Index.] | Gesetzlich deponirt. Verlag der Stenochromatischen Anstalt von Otto Radde, Hamburg. | [This work consists of a single separate leaf of unpaged text and 42 colored quarto plates, enclosed in book-shaped box.]

PART II.
ORNITHOLOGISTS' COMPENDIUM.

GLOSSARY OF TECHNICAL TERMS

USED IN

DESCRIPTIVE ORNITHOLOGY.

A.

Ab'domen (L *abdo'men*), n The belly (Plate XI.)
Abdo'minal, (L. *abdomina'lis*), a Pertaining to the abdomen.
Aber'rant, (L. *aber'rans*), a Deviating from the usual, or normal, character
Abnor'mal (L *abnorma'lis*), a Of very unusual or extraordinary character The opposite of normal.
Abor'tive (L *abor'tivus*), a. Imperfectly developed
Acces'sory, a. Joined to another thing; additional (as an accessory plume)
Accip'itres (L), n. Plural of *Accipiter*, also the name of a more or less artificial group of birds, including the so-called "Birds of Prey," or *Raptores* of some authors.
Accip'itrine (L *accipitri'nus*), a. Hawk-like.
Acic'ular (L *acicula'ris*), a. Needle-shaped. (Plate XIV fig 11.)
Acu'leate (L *aculea'tus*), a Slender-pointed
Acu'minate (L *acumina'tus*), a. Tapering gradually to a point.
Acute' (L *acu'tus*), a Sharp-pointed
Adoles'cence, n Youth.
Adult', n As applied to birds, an individual which has attained the final or mature plumage
Adult', a. In Ornithology, having reached the fully mature or final plumage. (A bird may be adult as regards organization without being of adult plumage)
Æstiv'al (L *æstiva'lis*), a. Pertaining to summer.
Aetomor'phæ (L), n. A name (signifying "eagle-formed") proposed by Professor Huxley for the Birds of Prey (*Raptores* or *Accipitres* of other authors)

Ag'gregated, *a* Collected together; accumulated Thus, by aggregation, a number of individual spots or other markings may form, collectively, a larger patch or stripe.

Affin'ed (L *af'finis*), *a* Related by affinity.

Affin'ity, *n* Direct relationship.

Af'ter-shaft, *n* Properly, the stem of the supplementary plume springing from near the base of some feathers; ordinarily, however, applied to the plume itself

Al'ar (L. *ala'ris*), *a*. Pertaining to the wing

Alaud'ine (L. *alaudi'nus*), *a*. Lark like.

Al'binism, *n* An abnormal condition of plumage, with white replacing the ordinary colors to a greater or less extent Albinism results from a deficiency or entire absence of pigment in the skin which supplies the coloring of the feathers, and is complete only when all colors are obliterated from the plumage (In birds, complete albinism of the plumage is not necessarily accompanied by change of colors of the bill, feet, and eyes)

Albi'no, *n* An animal affected with albinism

Albinis'tic,
Albinot'ic, } *a* Affected with albinism

Alec'troid (L *alectroi'deus*), *a*. Cock-like; resembling the domestic cock (*Gallus ferrugineus*, ♂)

Alec'torine (L *alectori'nus*), *a*. Pertaining to the domestic cock.

Alectoromorph'æ (L), *n* The Huxleyan name (meaning "cock-formed") for the Gallinaceous birds (*Gallineæ* or *Gallinaceæ* of other authors)

Al'iform (L *alifor'mis*), *a* Wing like

Alp'ine (L *alpi'nus*), *a* Pertaining to the Alps. (Often used in relation to any high mountain-range for species inhabiting high altitudes, which are termed "Alpine" species)

Al'trices (L), *n*. Birds whose young are reared in the nest and fed by the parents With the exception of the *Raptores*, some of the *Steganopodes* and *Pygopodes*, the *Longipennes* and *Sphenisci*, the young of the *Altrices* are psilopædic, that is, born naked, or only partially clad

Altri'cial, *a*. Having the character of, or pertaining to, the *Altrices*.

Alu'la (L; pl *alu'læ*), *n*. The "bastard-wing," composed of several stiff feathers growing on the so-called thumb They are situated directly below the secondary or greater coverts, and collectively resemble a miniature wing, whence the name (Plate XI.)

Alu'lar, *a*. Pertaining to the alula

Am'bulatory, *a* Gradient; walking or running (Opposite of *Saltatory*, hopping or leaping)

Amphimorph'æ (L), *n*. The Huxleyan name for a natural group, or so-called "order" of birds, including only the Flamingoes (*Phœnicopteridæ*).

A'nal (L. *ana'lis*), *a*. Pertaining to the anus

A'nal region (L. *re'gio-ana'lis*), *n.* The feathers immediately surrounding the anus. (Plate XI)

Analog'ical, *a.* Having analogy.

An'alogue, *n.* Anything having analogy with another. Thus, the *Cathartidæ* are the New World analogues of the Old World vultures

Anal'ogous, *a* Having analogy

Anal'ogy, *n.* Superficial or general resemblance, without structural agreement, or affinity, the resemblance between the *Cathartidæ* (New World Vultures) and the Vultures of the Old World (*Falconidæ*) is purely one of analogy.

Anal'ysis, *n.* In the usual natural history sense, the definition of species or higher groups by a tabular arrangement of characters, usually antithetical, with subdivisions under appropriate headings

Analyt'ic,
Analyt'ical, } *a.* An *analytical table* is a tabular arrangement of antithetical characters, distinguishing genera, species, or higher groups

An'atine (L *anati'nus*), *a* Duck-like

Ancip'ital, *a* Two-edged; double-edged.

Angle of Chin (L *an'gulus menta'lis*), *n* The anterior point of the space between the rhami of the lower jaw (See *Mental Apex* Plate XII fig 4)

Anisodac'tylæ (L), *n.* The name of a group of birds having three toes in front and one behind

Anisodac'tylous (L *anisodac'tylus*), *a.* Having three toes in front and one behind.

Anisopo'gonous, *a.* Said of a feather when the two webs are of unequal breadth

An'notine (L *annoti'nus*), *n.* A bird one year or less old, or which has moulted but once (Little used)

An'nular (L *annula'ris*), *a.* Ringed

Anomalogona'tæ (L), *n.* A primary subdivision of the order *Euripidura*, proposed by A. H. Garrod

Ano'malous (L *ano'malus*), *a.* Very unusual, strange; abnormal

An'serine (L. *anseri'nus*), *a* Goose-like

An'te (in composition) Anterior to, or before; as *anteorbital*, *anteocular*, etc

Ante'rior, *a* Forward, in front of.

An'thine (L *anthi'nus*), *a* Pipit like

Antith'esis, *n* An opposition of words or sentences distinguishing at a glance the diagnostic characters of two or more groups or species.

Antithet'ic,
Antithet'ical, } *a* Contrasted by, or pertaining to, antitheses

Antrorse', *a* Directed forward, as the nasal tufts of most jays and crows, and the rictal bristles of many birds.

Ant'werp Blue, *n* A very rich and intense blue color, similar to but purer than Prussian Blue (Plate IX fig 10)

A'pex (L, pl *a'pices*), *n.* The tip or point of anything

Apple Green, *n* A very light yellowish green color (Light green zinnober + lemon yellow + white) (Plate X fig 20)

Aquat′ic (L *aqua′ticus*), *a.* Pertaining to the water. Aquatic birds are those which derive their subsistence chiefly from seas, lakes, or rivers, and include two artificial groups known as "Waders" and "Swimmers."

Aq′uiline (L. *aquili′nus*), *a.* Eagle-like.

Arbo′real,
Arbor′icole, } (L *arbo′reus*), *a.* Tree-inhabiting.

Arcu′ate (L. *arcua′tus*), *a.* Bow-shaped; arched

Are′olæ (L. pl of *are′ola*), *n.* The small naked spaces between the scales of the feet, usually called *interspaces*

Arie′tiform (L *arie′tiform′is*), *a.* Having the form of the zodiacal sign *Aries,* ♈. (Plate XV. fig 11)

Armil′la (L), *n.* A colored ring round the lower end of the tibia; an anklet

Articula′tion, *n* A joint or hinge (Usually applied to the limbs)

Ash-color (L *cine′reus*), *n.* (See *Cinereous*) (Plate II fig 16)

As′ter Purple, *n* A rich clear purple color, like some varieties of the aster (Winsor & Newton's intense blue, or Schoenfeld's violet madder lake + Bourgeois's "rose tyrien") (Plate VIII fig 8.)

Asymmet′rical, *a* Without symmetry, or without close resemblance between corresponding parts, as opposite sides (The pattern of coloration in partial albinos is often asymmetrical)

Asym′metry, *n.* Disproportion, or want of close resemblance, between corresponding parts or organs (Very decided asymmetry of opposite sides of the skull is observable in some Owls)

At′rophy, *n* The wasting away, or obliteration, of an organ or part through deficient nutrition

Atten′uate (L *attenua′tus*), *a.* Tapering or growing gradually narrower toward the extremity, but not necessarily pointed (which would be *acuminate*).

Auric′ular (L *auricula′ris*), *a* Pertaining to the ear

Auric′ula Purple, *n* A deep but rather dull purple, like the color of the purple auricula (Schoenfeld's violet madder lake, or Winsor & Newton's violet carmine) (Plate VIII fig 3)

Auric′ulars (L *re′gio auricula′ris*), *n* The (usually) well-defined feathered area which conceals the ears in birds (Plate XI)

Autop′tical, *a* Personally inspected.

Autum′nal Plu′mage (L *ves′tis autumna′lis*), *n.* The full dress of autumn In most birds it remains essentially unchanged till the spring moult In many species the young possess a peculiar autumnal plumage (assumed by their first moult) which differs not only from their first livery but also from that of adults at the same season In such, the adult or mature plumage may be completely assumed at the next moult, or it may be gradually acquired by successive moults, as in the case of many Orioles (*Icteridæ*), Tanagers, and other bright-colored Passerine groups.

Av'ian Fauna, } *a* The bird-life of a particular country or locality.
Avi-fauna,

Ax'illa (L.), *n.* The armpit.

Ax'illar, } (L *axilla'ris*), *a.* Pertaining to the armpit.
Ax'illary,

Ax'illaries, } (L *axilla'res*), *n.* A more or less distinct tuft of graduated, usually soft and elongated, feathers growing from the armpit. (Plate XIII. fig. 4.)
Ax'illars,

Az'ure Blue (L. *azu'reus*), *n* A fine light blue color, like the blue of the sky. (Cobalt blue + white) (Plate IX. fig 15.)

B.

Back (L. *dor'sum*), *n.* In descriptive Ornithology, usually includes the scapulars and interscapulars, but should properly be restricted to the latter alone. (Plate XI.)

Band (L *vit'ta; fas'cia*), *n.* A broad transverse mark with regular and nearly parallel edges, a broad bar of color (A broad band is usually called a *zone*) (Plate IV. fig. 18.)

Band'ed (L *vitta'tus, fascia'tus*), *a.* Marked with bands. (Plate XV. fig 18)

Bar (L *vit'ta, fas'cia*), *n* A narrow transverse mark of color. (Plate IV fig. 17)

Barb (L *bar'bus*), *n.* Any one of the fibrillæ, or laminæ, composing the web of a feather

Barb'ed (L *barba'tus*), *a* Furnished with barbs, bearded

Barb'ule (L *bar'bulus*), *n.* A barb of a barb

Barred (L *vitta'tus; fascia'tus*), *a* Marked with bars (Plate XV. fig 17)

Base (L *ba'sis*), *n* Root; origin

Ba'sal (L *basa'lis*), *a* Pertaining to the base

Bay (L. *ba'dius*), *n* A very rich dark reddish chestnut. (Burnt sienna + purple madder) (Plate IV fig 5)

Bel'ly (L *abdo'men*), *n* The central posterior portion of the under surface of the body, bounded laterally by the sides, posteriorly by the vent or anal region, and anteriorly by the breast (Plate XI)

Belt (L *bal'teus*), *n* A broad band of color across the breast or belly (Distinguished from *zone* in that the latter may cross the wings or tail)

Belt'ed (L *baltea'tus*), *a.* Marked with a broad band or belt of color across the lower part of the body.

Bend of the Wing (L *flex'ura; pli'ca*), *n* The angle or prominence at the carpus, or wrist-joint, in the folded wing (Plate XIII fig 5)

Berlin' Blue, *n.* A deep dark blue color, rather lighter and less purplish than marine blue. (Schoenfeld's Berlin blue.) (Plate IX fig 4.)

Beryl Green (L. *beryli'nus*), *n.* A light bluish green similar to verdigris, but more bluish. (Dark permanent green + Schoenfeld's "licht blau") (Plate X fig 14.)

Bev'elled, *a.* Having two plane surfaces joining obliquely.

Bev'y, *n.* A flock of quails or partridges.

Bi- (in composition). Twice, double. As *bicolored* (two-colored), *biped* (two-footed), *bifurcate* (double-forked), etc.

Bibliog'raphy, *n.* Condensed history of the literature of a subject.

Bice Green, *n.* A yellowish green color, lighter and more yellow than parrot green. (Light zinnober-green + lemon-yellow.) (Plate X. fig. 10.)

Bi'colored (L. *bi'color*), *a.* Two-colored.

Bifur'cate (L. *bifurca'tus*), *a.* Doubly forked.

Bino'mial,
Bino'minal, *a.* Two-named, or, more properly, named by two terms. The *binomial system of nomenclature*, instituted in 1758 by Linnæus, and adopted by zoologists and botanists, promulgates the use of two terms as the name of each species, — the first generic, the second specific.

Biol'ogy, *n.* The study of living beings with relation to the laws and results of their organization.

Biolog'ical, *a.* Pertaining to Biology. *Biological science* embraces the study of all organic creations, and thus includes Zoology and Botany, both recent and fossil.

Bis'tre, *n.* A dark brown color somewhat more reddish than sepia, but much less so than burnt umber. (Plate III. fig 6.)

Boat-shaped (L. *cymbifor'mis*), *a.* A *boat-shaped tail* has the opposite sides, or halves, *meeting below* along the median line, the outer edges being elevated. The tail of *Quiscalus* (Boat-tailed Blackbird) is a familiar example, while that of the domestic fowl (*Gallus bankiva*) exemplifies the opposite form, with the edges below and the middle feathers forming the *ridge* instead of the keel. A *boat-shaped bill* is one in which the maxilla resembles an inverted boat, as in the genus *Cancroma*.

Boot, *n.* In birds, the tarsal envelope, when entire.

Boot'ed (L. *ocrea'tus*), *a.* A booted tarsus has the usual scales fused so as to form a continuous or uninterrupted covering. The tarsus of the smaller Thrushes and the American Robin (*Merula migratoria*) well illustrates this character.

Bor'dered (with) (L. *limbatus*), *a.* Having the edge or margin all round of a different color.

Bo'real (L. *borea'lis*), *a.* Northern.

Boss, *n.* A knob or short rounded protuberance.

Bottle Green, *n.* A dark green color like the color of some varieties of glass. (Schoenfeld's "dark zinnober-green" or Winsor & Newton's Prussian green + Winsor & Newton's "olive-green") (Plate X fig 1.)

Brac'cate (L *bracca'tus*), *a.* Having the feathers on the outer side of the tibia elongated, or plume-like, as in most of the Falconidæ

Brace-shaped, *n.* Shaped like the brace (⌢⌢) of printers (Plate XV. fig. 6.)

Brach'ial (L *brachia'lis*), *a.* Pertaining to the wing (Little used.)

Brachyp'terous (L *brachyp'terus*), *a.* Short-winged

Brachyu'rous (L *brachyu'rus*), *a.* Short-tailed.

Breast (L *pec'tus*), *n.* In birds, an artificial and somewhat arbitrary subdivision of the under surface, lying between the jugulum and abdomen Its position corresponds nearly with that of the underlying pectoral muscles (Plate XI.)

Brevipen'nes (L.), *n.* The systematic name of a group of short-winged birds, including the ostriches and kindred forms

Brevipen'nine (L *brevipen'nis*), *a.* Short feathered; short-winged (improperly so used), pertaining to the *Brevipennes*

Brick Red (L *testa'ceus, lateri'tius; ru'tilus*), *n.* A dull brownish red color like the color of burnt bricks. (Corresponds very nearly with Winsor & Newton's Indian red Same as tile red) (Plate IV fig 11)

Bridle (L *fre'num*), *n.* A stripe of color extending back from the bill, along the lower sides of the head

Bridled (L *frena'tus*), *a.* Marked with a distinct stripe of color from the bill backward, beneath the eye, along the lower jaw, or the sides of the throat

Bris'tle, *n.* A small hair-like feather, consisting chiefly of the shaft, commonly developed near the angle of the mouth, or rictus, but sometimes on other portions of the plumage also

Broc'coli Brown, *n.* A grayish brown color, intermediate in tone between drab and hair-brown. (Bistre + raw umber + black + white.) (Plate III fig 15)

Buc'cal (L *bucca'lis*), *a.* Pertaining to the cheeks

Buff (L *lu'teus, luteo'lus*), *n.* A light dull brownish yellow, like the color of dressed buckskin or chamois (Raw sienna + white.) Plate V fig 13

Buff-Pink, *n.* A pink color tinged with, or inclining to, buff. (Light red + cadmium orange + white) (Plate IV fig 20)

Buff-Yellow (L *lu'teo-fla'vus*), *n.* A yellow color tinged with or inclining to buff (Orange-cadmium + pale cadmium + white) (Plate VI fig 19)

Bul'late (L *bulla'tus*), *a.* Having a blistered appearance

Burnt Carmine, *n.* A very rich brownish crimson inclining to maroon or claret-color (Madder-carmine + scarlet-vermilion + black) (Plate VII. fig 1)

Burnt Sienna (L *spadi'ceus*), *n.* A rich reddish brown color, like the pigment of the same name (Plate IV fig 6)

Burnt Umber (L *satura'te umbri'nus*), *n.* A deep rich brown color, more reddish than sepia and bistre. (Plate III fig 8)

C.

Cad′mium Orange (L *cadmium′no auran′tius*), *n* An exquisitely rich, mellow orange-color (much purer in tint than a true orange) like the breast of the fully adult male Baltimore oriole, or the throat of the Blackburnian Warbler (Plate VI fig 2)

Cad′mium Yellow (L. *cadmium′no-fla′tus*), *n* A very intense pure orange-yellow color, a little deeper and much purer than Indian yellow, and much more orange than gamboge. (Plate VI fig 6)

Cadu′cous (L *cadu′cus*), *a* Falling off early.

Calca′reous, *a* Chalky

Campan′ula Blue, *n.* A moderately deep purplish blue color, like the hue of some species or varieties of the blue-bell or Canterbury-bell (*Campanula*) (Smalt + white) (Plate IX. fig. 11)

Canal′iculated (L *canalicula′tus*), *a* Channelled or furrowed

Cana′ry Yellow, *n* A delicate pure yellow color, paler than gamboge but deeper than maize or primrose Nearly the same tint as King's yellow. (Schoenfeld's "heller cadmium" and white) (Plate VI. fig 12)

Can′cellate (L. *cancella′tus*), *a.* Latticed, marked both longitudinally and transversely

Candes′cent (L *candes′cens*), } *a.* Whitish; hoary; frost like.
Canes′cent (L *canes′cens*), }

Cap′illary (L *capilla′ris*), *a* Hair-like

Cap′istrate (L *capistra′tus*), *a* Hooded or cowled.

Capis′trum (L), *n* A hood or cowl In descriptive Ornithology, the fore part of the head all round, or that portion immediately surrounding the base of the bill

Cap′ital (L *capita′lis*), *a* Pertaining to the head.

Cap′itate (L *capita′tus*), *a* A *capitate feather* has the end enlarged.

Cap′ut (L , gen *cap′itis*, pl *cap′ita*), *n* The head.

Carbona′ceous (L *carbona′ceus*), *a* Pertaining to carbon, or charcoal Thus, *carbonaceous-black* = coal-black

Cari′na (L) *n* A keel, or median ridge

Car′inate (L *carina′tus*), *a* Keeled, or with a median ridge *Carinate Birds* (*Aves carinatæ*) are those furnished with a keeled sternum

Car′neous (L. *car′neus*), *a* Fleshy

Carniv′orous (L *carniv′orus*), *a* Flesh-eating

Car′mine (L *carmin′eus; coccin′eus*), *n* A very pure and intense crimson The purest of the cochineal colors. (Madder-carmine + scarlet-vermilion) (Plate VII fig 6)

Car′pal (L. *carpa′lis*), *a.* Pertaining to the wrist, or carpus

Car'pal joint, Car'pal angle, { (*L carpus*), *n.* The prominence formed by the wrist-joint, or carpus, when the wing is closed. The length of the wing, in descriptions, is measured from the carpal angle to the tip of the longest quill. (Same as *Bend of the Wing*, or *flexura*.) (Pl XIII fig. 5.)

Car'po-metacar'pal joint, *n* The last wing-joint, covered exteriorly by the alula. (Plate XIII fig 6.)

Car'pus (L.), *n.* The wrist. In a bird, the space between the bend (*flexura*) and the hand-joint of the wing.

Car'uncle (L *caruncalus*), *n.* A naked fleshy excrescence, usually about the head or neck, and ordinarily brightly colored, wrinkled, or warty.

Carun'culate, Carun'culated, } (L *caruncula'tus*), *a* Having caruncles.

Castan'eous (L. *casta'neus*), *a* or *n* Chestnut-colored, chestnut-color (Plate IV. fig 9.)

Caud'al (L *cauda'lis*), *a* Pertaining to the tail.

Caud'a (L.), *n* The tail.

Caud'ate (L *cauda'tus*), *a* Tailed.

Cecomorph'æ (L.), *n.* The Huxleyan name (meaning "gull-formed") of the gull-tribe, or *Gaviæ*.

Celeomorph'æ (L.) *n.* The Huxleyan name (meaning "woodpecker-formed") of the natural group of birds including the Woodpeckers (*Picidæ*).

Cephal'ic (L. *cephal'icus*), *a.* Pertaining to the head.

Cera'ceous (L. *cera'ceus*) *a.* Wax-like

Cere (L *ce'ra*), *n.* The naked skin or membrane in which the nostrils are situated, common to most Birds of Prey (*Raptores*) and many of the Parrot-tribe (*Psittaci*), as well as the Pigeons (*Columbæ*) and some other groups. It usually has a more or less distinct line of demarcation anteriorly (except in the Pigeons)

Ceru'lean Blue (L *cæru'leus; cælesti'nus, cæles'tis; cælico'lor*), *n.* A fine light blue color, less purplish or more greenish than azure (Winsor & Newton's cerulean blue.) (Plate IX fig 21.)

Cer'vical (L *cervica'lis*), *a.* Pertaining to the cervix or hind neck.

Cer'vix (L.), *n* The hind neck, extending from the occiput to the commencement of the back. It has two subdivisions, namely, the nape and scruff (*nucha* and *auchenium*), which occupy respectively the upper and lower halves of the cervix. (Plate XI.)

Change'able, *a* As applied to colors, varying in tint with different inclinations to the light iridescent

Char'acter, *n.* Any peculiarity of structure or plumage, or other distinctive attribute, available for the diagnosis of a species, genus, or higher group.

Charadriomorph'æ (L.), *n* The Huxleyan name (meaning "plover-formed") for the group of smaller wading birds usually called *Limicolæ*

Cheek (L *ge'na, buc'ca*), *n* An arbitrary subdivision of the side of the head, differently employed by various writers, but usually corresponding to the *malar region*, or the feathered portion of the lower jaw. (Plate XII.)

Chenomorph'æ (L.), *n.* The Huxleyan name (meaning "goose-formed") of a group of birds equivalent to the *Anatidæ* of authors.

Chest'nut (L. *casta'neus; spadi'ceus*), *n.* A rich dark reddish brown, of a slightly purplish cast. (Vermilion + burnt umber.) (Plate IV. fig. 9.)

Chin (L. *men'tum*), *n.* The extreme anterior point of the gular region, or the space between the lateral branches (*rhami*) of the lower jaw. (Plate XI.)

China Blue, *n.* A dull medium blue color. (Intense blue + white.) (Plate IX. fig. 13.)

Chinese Or'ange, *n.* A very intense orange-red color, of a peculiar tint, very different from orange-chrome. (Cadmium-orange + burnt sienna.) (Plate VII fig. 15.)

Choc'olate Brown (L. *chocolati'nus*), *n.* A rich dark reddish brown color, like the exterior glazed surface of a cake of chocolate. (Purple madder + sepia.) (Plate III fig. 4.)

Chrome Yellow, *n.* A deep yellow, much less pure or intense than light cadmium. (Winsor & Newton's "chrome-yellow.") (Plate VI fig. 8.)

Chro'mium Green, *n.* A dull green color, nearly intermediate between malachite green and sage green. (Green oxide of chromium.) (Plate X fig. 12.)

Cic'onine (L. *ciconi'nus*), *a.* Stork-like.

Cil'ium (L., pl. *cilia*), *n.* An eyelash.

Cine'reous (L. *cine'reus*),
Cinera'ceous (L. *cinera'ceus*), } *n.* Ash-gray, a clear bluish gray color, lighter than plumbeous. (Lamp-black + Chinese white.) (Plate II fig 16.)

Cinnamo'meous (L. *cinnamo'meus*),
Cin'namon (L. *cinnamomi'nus*), } *n.* or *a.* A light reddish brown color, like the inner surface of cinnamon bark. (Indian red + raw umber.) (Plate III fig 20.)

Cin'namon Ru'fous (L. *cinnamo'meo-rufus*), *n.* Rufous, with a tinge of cinnamon. (Burnt sienna + burnt umber + light red + white.) (Plate IV. fig 16.)

Cir'cular, *n.* Of a rounded shape. (Plate XIV fig. 4.)

Cir'cum- (in composition) Around, encircling, as, *circumorbital* (around the eye), *circumventral* (around the vent), etc.

Cir'rhous (L. *cirra'tus*), *a.* Tufted.

Cit'ron Yellow, *n.* A light greenish yellow, deeper and less pure than sulphur-yellow. (Light cadmium + light zinnober-green.) (Plate VI. fig 15.)

Clar'et Brown (L. *vina'ceo-brunn'eus*), *n.* A rich dark brownish purple, much like the pigment called "Purple-madder." Nearly the same as "maroon," but more purple. (Purple-madder.) (Plate IV. fig. 1.)

Class (L. *clas'sis*), *n.* A primary division of animals, as the *class* of Birds (*Class Aves*).

Classifica'tion, *n.* A systematic arrangement.

Claw (L. *un'guis*), *n.* The horny, pointed, and compressed sheath of the terminal phalanx of the toe.

Clay-color (L *lutes'cens, lute'olus, luto'sus*), *n.* A dull light brownish yellow color, nearly intermediate between yellow-ochre and Isabella-color (Yellow ochre + raw umber + white) (Plate V fig 8)

Clove Brown, *n* A dark brown color, like dried cloves (Black + cadmium-orange). (Plate III. fig 2.)

Clutch, *n.* A nest-complement or "set" of eggs

Co'balt Blue (L. *cobalti'nus*), *n* A very fine pure light blue color, less intense and more azure than ultramarine (Plate IX. fig 12)

Coccy'ges (L) *n* The systematic name of a natural group of zygodactyle birds, including the Cuckoos (*Cuculidæ*), Plantain-eaters, Turacous (*Musophagidæ*), Trogons (*Trojonidæ*), etc

Coccygomorph'æ (L) *n* The Huxleyan name (meaning "Cuckoo-formed") of the *Coccyges*

Col'lar (L *tor'ques*), *n.* A ring of color encircling the neck

Col'lared (L *torqua'tus, colla'ris*), *a* Marked with a neck-ring of a different color from surrounding parts

Col'lum (L), *n.* The neck

Colora'tion, *n.* Pattern of coloring, or the colors of the plumage collectively.

Col'ored (L *colora'tus*) *a* In Ornithology, different from white. Thus, the *colored phase* of a dichromatic species is that in which the plumage is other than white

Comb, *n* An erect, fleshy, longitudinal caruncle on the top of the head, as in the domestic fowl (*Gallus ferrugineus*, var) and the adult male Condor (*Sarcorhamphus gryphus*).

Commis'sural, *a* Pertaining to the commissure

Com'missure (L. *commis'sura*), *n* The outlines of the closed mouth, or the opposed edges of the mandible and maxilla.

Compress'ed, *a* Flattened sideways, or higher than broad A *compressed tail* has the two halves folded together with the two edges separated below, the median feathers forming the ridge, as in the domestic fowl. (The opposite form is seen in the *boat-shaped* tail of the American Grackles, *Quiscalus*)

Con'cave (L *conca'vus*), *a* Hollowed on one side, as the inside of a curved line, the under side of an arch, or the hollow of a spoon

Concen'tric (L *concen'tricus*), *a*. Having a common centre, as a series of rings one within another. (Plate XV. fig 19)

Con'colored (L *conco'lor*), *a*. Of a uniform color (Same as *unicolored*)

Con'fluent (L *confluen'tus*), *a* Run together

Con'gener, *n* A species belonging to the same genus with another.

Congener'ic, *a.* Belonging to the same genus with another

Coniros'tral, *a* Having a conical bill, like that of a Finch or Sparrow, pertaining to the so-called *Conirostres*.

Coniros'tres (L), *n* An arbitrary group of birds, in classifications, of which the Sparrow tribe (*Fringillidæ*) are typical

Contin'uous, *a* (As applied to markings.) Without interruption.

Con'tour Feathers, *n.* The surface feathers of the head, neck, and body

Coracomorph'æ (L.), *n.* The Huxleyan name (meaning raven-formed) for the Passeres.

Cor'al Red (L *coralli'nus, coralli'no-ru'ber*), *n.* A light, rather dull vermilion, like the color of red coral. (Madder-carmine, orange-vermilion + white.) (Plate VII fig 4.)

Cord'ate (L *corda'tus*), } *a.* Heart-shaped. (Plate XIV fig 15.)
Cord'iform (L *cordifor'mis*), }

Coria'ceous (L *coria'ceus*), *a.* Of leathery texture.

Corn'eous (L. *cor'neus*), *a.* Horny.

Cornic'ulate (L *cornicula'tus*), *a.* Furnished with a small horn.

Corn'iplume, *n.* A horn-like tuft of feathers on the head.

Cor'onate (L *corona'tus*), *a.* Crowned, having the top of the head ornamented by lengthened or otherwise distinguished feathers.

Cor'rugate, } (L. *corruga'tus*), *a.* Wrinkled.
Cor'rugated, }

Co'vey, *n.* A family (or brood with or without their parents) of Quails or other game-birds.

Cream-color, *n.* A light pinkish yellow color, like cream. (Cadmium yellow + white.) (Plate VI fig 20.)

Cream'y Buff, *n.* (Yellow ochre + white.) (Plate. V. fig 11.)

Cre'nate, } (L *crena'tus*), *a.* Having rounded teeth. (Plate XV fig 21.)
Cre'nated, }

Cren'ulate (L *crenula'tus*), *a.* Finely crenate.

Crepus'cular (L *crepuscula'ris*), *a.* Pertaining to twilight. (Crepuscular birds are those which become active after sunset.)

Cres'cent, *n.* A figure having the shape of the new moon.

Crescent'ic (L *luna'tus*), *a.* Shaped like the new moon. (Plate XV fig. 9.)

Crest (L *cris'ta*), *n.* A more or less lengthened, erectile, or permanently erect, tuft of feathers on top of the head.

Crest'ed (L *crista'tus*), *a.* Furnished with a crest.

Crim'son (L *carmesi'nus, sanguin'eus; sanguin'eo-ru'ber*), *n.* Blood-red, the color of the cruder sorts of carmine. (Madder-carmine, or dark madder-lake.) (Plate VII fig 3.)

Cris'sum (L.), *n.* A term usually applied to the lower tail-coverts collectively, but properly belonging to the feathers situated between the lower tail-coverts and the anal region. (Plate XI. See especially note facing plate.)

Cris'sal (L *crissa'lis*), *a.* Pertaining to the crissum.

Crown (L *coro'na*), *a.* Properly the vertex, or that portion of the top of the head between the forehead and the occiput. (Plate XI.)

Cru'ciate (L *crucia'tus*), } *a.* Cross-like. (Plate XV fig 10.)
Cruci'form (L *crucifor'mis*), }

Cru'ral (L *crura'lis*), *a.* Pertaining to the crus, or tibia.

Crus (L.), *n.* The "thigh," or tibia.

Cu'bital (L *cubita'lis*), *a* Pertaining to the forearm

Cu'cullate (L. *cuculla'tus*), *a* Hooded, or having the head colored differently from the rest of the plumage

Cul'men (L), *n* The ridge or upper outline of the maxilla, or upper mandible. (Plate XII fig 7)

Cul'minal (L. *culmina'tus*), *a* Pertaining to the culmen

Cul'trate (L *cultra'tus*), *a* Knife-like

Cultriros'tral (L *cultriros'tris*), *a* Having a knife-shaped bill, or the bill lengthened, compressed, and pointed, like a Heron's Pertaining to the so-called *Cultrirostres*.

Cultriros'tres (L.), *n* An artificial group of wading birds, including the Herons (*Ardeidæ*) and Storks (*Ciconiidæ*), so named on account of the knife-shaped bill, and in this sense nearly equivalent to *Herodiones*. Also applied to a group of Passerine birds, which includes the *Corvidæ, Sturnidæ, Icteridæ,* etc.

Cu'neate (L *cunea'tus*), } *a*. Wedge-shaped. (Plate XIV. fig 13)
Cu'neiform (L *cuneifor'mis*), }

Cu'preous (L *cu'preus*), *a*. Coppery, like copper

Curso'res (L), *n* An artificial group of birds, in the older systems, including the Bustards and other "coursers" or "runners"

Curso'rial, *a*. Running, pertaining to the *Cursores*.

Cus'pidate (L. *cuspida'tus*), *a*. Stiff-pointed

Cuta'neous, *a* Pertaining to the skin (Same as *dermal*)

Cyp'seline (L *cypseli'nus*), *a*. Swift-like, pertaining to the *Cypselidæ*, or Swifts

Cypselomorph'æ (L), *n* The Huxleyan name (meaning "swift-formed") of a group of ægithognathous "*Picariæ*," including the Goat-suckers (*Caprimulgidæ*), Swifts (*Cypselidæ*), and Humming-birds (*Trochilidæ*) (Equivalent to the *Macrochires* and *Cypseli* of other authors)

Cylin'dric-o'vate, *n* An elongate ovate with parallel sides. (Plate XVI fig. 6)

Cym'biform (L *cymbifor'mis*), *a*. Boat-shaped

D.

Dah'lia Pur'ple, *n* A rich dark purple color, like some varieties of the dahlia (*Dahlia variabilis*). (Madder-carmine + intense blue) (Plate VIII fig 2)

Dasypæ'dic, *a* Clothed with down at birth (Same as *Ptilopædic*)

Decid'uous, *a*. Temporary, or shed periodically, as the horns of a deer and the " nuptial ornaments" of many birds

Declin'ate, }
Declin'ed, } (L *declina'tus*), *a* Bent downward.

Decompos'ed, *a* Said of a feather when the barbs are separated, not forming a continuous or compact web

Decum'bent (L *decum'bens*), *a.* Hanging downward, drooping

Decus'sate (L *decussa'tus*), *a* Crossed, intersected (Plate XV fig 16)

Deep Chrome Yellow, *n.* A very deep but not brilliant yellow color. (Winsor & Newton's "deep chrome," or Schoenfeld's "goldgelb" or "mittel chromgelb") (Plate VI fig 9)

Del'toid (L *deltoid'eus*), *a* Triangular, or shaped like the Greek character "Delta," Δ. (Plate XIV fig 14)

Den'tate (L *denta'tus*), *a.* Toothed (Plate XV fig 20)

Dentic'ulate (L. *denticula'tus*), *a.* With small teeth

Dentig'erous (L *dentiger'*), *a* Bearing teeth

Dentiros'tres (L), *n.* An artificial or arbitrary group in classifications, the members of which have the maxilla more or less notched near the tip

Dentiros'tral (L *dentirostris*), *a* Tooth-billed, pertaining to the *Dentirostres*

Denuda'tion, *a* Nakedness

Deplum'ate (L *depluma'tus*), *a.* Bare of feathers

Depress'ed (L *depres'sus*), *a.* Flattened vertically; broader than high (Opposite of compressed)

Der'mal (L *derma'lis*), *a.* Pertaining to the skin.

Desquama'tion, *n.* Peeling or scaling off

Di- (in composition) Twice, double (as *dichromatic* = two colored).

Diagno'sis, *n* A condensed statement of the characters which are exclusively applicable to a species, genus, or higher group, a description which omits all non essential characters

Diagnos'tic, *a* Pertaining to diagnoses; exclusively applicable, or distinctly characteristic

Dichot'omous, *a* Paired, or by twos.

Dichromat'ic, *a* In descriptive Ornithology a species is said to be *dichromatic* when it exists in two distinct plumages which are entirely independent of sex, age, or season. These distinct plumages were formerly, in the case of most dichromatic birds, supposed to represent distinct species, and the nature of their real relationship is a comparatively recent discovery. Familiar examples of dichromatism are the rufous and gray forms of the little Screech Owl (*Scops asio*), and the white and bluish or dusky forms of some Herons (as *Ardea occidentalis* and *Dichromanassa rufa*).

Dichrom'atism, *n.* The state of existing in two distinct phases of coloration, which are wholly independent of the usual causes of color differences (as sex, age, and season) Dichromatism among birds is somewhat analogous to *dimorphism* in insects.

Didac'tyle,
Didac'tylous, } (L *didac'tylus*), *a.* Two-toed, as the Ostrich.

Dig'itigrade, *a* Walking on the toes (Applicable to most birds.)

Dimorph'ic, *a* Existing in two forms, as some species of insects

Dimorph'ism, *n* The state of existing in two forms (The nearest approach to dimorphism among birds is the state of *dichromatism*, which see)

Disc, } *n* Set of radiating feathers surrounding the eye in some birds,
Disk, } especially the Owls

Dis'tal, *a* Toward or at the extremity (Opposite of *proximal*, or toward the base)

Dis'tichous (L *dis'tichus*), *a* Two-rowed, as the webs of a feather.

Ditok'ous, *a* Producing but two eggs for each clutch, as the Pigeons (*Columbidæ*), Humming-birds (*Trochilidæ*), and a few other groups

Diur'nal (L *diurna'lis*), *a* Pertaining to the daytime Among birds, those which are active during the daytime and repose at night. (Many diurnal birds, however, are *nocturnal* in their migrations).

Divar'icate (L *divarica'tus*), *a* Spreading or curving apart

Dor'sal (L *dorsu'lis*), *a* Pertaining to the back

Dor'sum (L), *n*. The back. (Plate XI.)

Double-emarginate, } (L *biemargina'tus*), *a*. A *doubly emarginate tail*
Doubly emarginate, } has the middle and lateral feathers slightly longer than the intervening ones

Double-forked, } (L *bifurca'tus*), *a* A *doubly forked tail* has the middle
Doubly forked, } and lateral feathers decidedly longer than those between

Double-rounded, } (L *birotunda'tus*), *a*. A *doubly rounded tail* has the
Doubly rounded, } middle and lateral feathers shorter than those between

Down (L *floc'cus*), *n* Small soft decomposed feathers, which clothe the nestlings of many birds, and which also grow between and underneath the true feathers in the adults of many others, especially the various kinds of water-fowl

Down'y (L *pubes'cens*), *a*. Pertaining to or having the nature of down, or clad with down

Drab, *n* A brownish gray color. (Black + white + raw umber) (Plate III fig 18)

Drab-Gray, *n* (Black + white + burnt umber) (Plate II fig 13)

Dragon's-blood Red, *n*. A rich brownish red color, of a peculiar tint. (The pigment called *dragon's blood* is made from the inspissated juice of certain tropical plants, particularly the *Calamus draco* and *Dracæna draco*) (Light red + madder-brown.) (Plate IV fig 8)

Dusk'y (L. *obscu'rus*, *nigres'cens*, *m'gricans*), *n* or *a*. A dark color of more or less indefinite or neutral tint, ot a dark, indefinite color

Dysporomorph'æ (L), *n* The Huxleyan name (meaning gannet-formed) for the *Steganopodes*.

E.

Ear-cov′erts (L. *re′gio auricula′ris*), *n.* The usually well-defined tract of feathers overlying the ears of most birds. The ear coverts (or *auriculars*, as they are usually termed in descriptions) are bounded above by the backward extension of the *supercilium*, or lateral portion of the crown, posteriorly by the occiput and nape, below by the malar region or "cheeks," and anteriorly by the suborbital region. Same as *auriculars* (Plate XI.)

Ear′ed (L. *auri′tus*), *a.* Decorated with tufts of feathers, distinguished either by length or color, which by their appearance suggest the external ears of mammals

Ear-tufts, *n.* Erectile tufts of elongated feathers springing from each side of the crown or forehead, and presenting a close superficial resemblance to the external ears of many mammalia. They are especially characteristic of certain Owls (*Strigidœ*).

Econ′omy, *n.* Physiological disposition.

E′cru Drab, *n.* A very light, somewhat pinkish, drab color. (Burnt umber + sepia + white.) (Plate III. fig 21.)

Ec′to- (in composition) Outer, as *ectozoon*, an external parasite.

Edg′ed (with) (L *limba′tus*), *v* Having the edge or lateral margin of a different color

Edge of wing (L *campte′rium*, *mar′go-car′pi*), *n.* The anterior border of the wing, from the armpit to the base of the outer primary.

El′evated, *a* Said of the *hallux*, or hind toe, when inserted above the level of the anterior toes

Ellip′tical, *n* Having the form of an ellipse (Plate XIV. fig. 9, plate XVI fig 14.)

Ellip′tical-oval, *n.* See plate XVI fig 10

Ellip′tical-ovate, *n.* See plate XVI fig 4

Elon′gate (L. *elonga′tus*), *a* Lengthened.

Elon′gate-ovate, *n* See plate XVI fig 5.

Emarg′inate,
Emarg′inated, (L. *emargina′tus*), *a.* An *emarginate tail* has the middle feather shortest, the rest successively a little longer, hence an emarginate tail is very slightly forked An *emarginate quill* has the web suddenly narrowed by an abrupt cutting away of the edge. (Plate XIII. fig *a*.)

Em′bryo, *n* In birds, the young before leaving the egg

Em′erald Green (L *smaragdi′nus*), *n* A very bright light peculiar green color, like an emerald, but more especially like the pigment so called (Plate X fig 16.)

En′sate (L *ensa′tus*),
En′siform (L *ensifor′mis*), } *a.* Sword-shaped

En′to- (in composition). Inner; as *entozoön*, an internal parasite.

Ep'i- (in composition) Upon; as *epidermis*, upon the skin, that is, the surface skin.
Epider'mis (L), *n.* The cuticle, or scarf-skin.
Epider'mic, *a.* Pertaining to the epidermis.
Epignath'ous (L *epigna'thus*) Hook-billed, as a Hawk or Parrot.
Epithe'ma, *n.* A horny excrescence upon the bill.
Erec'tile, *a.* Capable of being raised or erected, as an *erectile crest*.
Erythris'mal, *a.* The state of being red or rufous, instead of the usual or "normal" color.
E'rythrism (L. *erythris'mus*), *n.* A particular state of plumage caused by excess of red or rufous pigment; it is one of the *dichromatic* states of many birds, as certain species of Owls (*Strix stridula, Scops asio*, etc.), also some species of *Accipiter* and *Micrastur*, among Hawks.
E'tiolated (L *œthiola'tus*), *a.* Whitened, bleached.
Etyp'.cal, *a.* Tending away from normal or typical character.
Eurhipidu'ra (L), *n* The name of one of the primary groups of birds, comprising *all existing species.*
E'ven (L *trunca'tus*), *a.* An even or "square" tail has, *when closed,* all the feathers terminating on the same transverse line; in other words, it is *truncated* at the tip. When spread, the tips of the feathers describe a semicircle, while an emarginate or slightly forked tail becomes even or truncated when spread.
Ex- (in composition) Out; out of; away from. As, *exterior*, on the outside.
Excres'cence, *n.* Any outgrowth, whether cutaneous, corneous, or fleshy.
Exot'ic (L *exot'icus*), *a.* Foreign.
Exten'sile (L *exten'silis*), *a.* Susceptible of being extended or lengthened.
Eye'brow (L *supercil'ium*), *n.* The middle portion of the superciliary region, or that part immediately above the eye.
Ey're,
Ey'rie, } *n.* The nest of a bird of prey, especially an Eagle.

F.

Fa'cial (L *facia'lis*), *a.* Pertaining to the face.
Fal'cate (L *falca'tus*),
Fal'ciform (L *falcifor'mis*), } *a.* Shaped like a sickle or scythe.
Fal'conine (L. *falconi'nus*), *a.* Falcon-like.
Fam'ily (L. *fami'lia*), *n* A systematic group in scientific classification, embracing a greater or less number of genera which agree in certain characters not shared by other birds of the same Order. In rank, a *Family* stands between *Order* and *Genus*, the former being composed of a greater or less number of nearly related families. In

zoological nomenclature the name of a Family is taken from a typical Genus, the name of which is modified by the termination *idœ*, as *Falconidœ, Columbidœ*, etc. (*Subfamilies* are distinguished by the termination *inœ*.)

Fas′cia (L.), *n.* A band or broad bar of color

Fas′ciated (L. *fascia′tus*), *a.* Banded or broadly barred

Fas′cicle (L. *fasci′culus*), *n.* A bundle

Fas′cicled, } (L. *fascicula′tus*), *a.* Bundled
Fascic′ulate,

Fastig′iate (L. *fastigia′tus*), *a.* Bundled together like a sheaf.

Fau′na (L.), *n.* The animal-life of a country. (Distinguished from the *Flora*, or plant-life.)

Fawn-color, (L. *cervinus; cervin′eus*), *n.* A light warm brown color. (Burnt umber + white.) (Plate III fig. 22.)

Fem′oral (L. *femora′lis*), *a.* Pertaining to the thigh proper, or the inner segment of the leg. (To be carefully distinguished from *tibial*, which refers to the so-called "thigh," or middle segment of the leg.)

Fe′mur (L.), *n.* The thigh, the thigh-bone.

Fe′ral (L. *fe′rus*), *a.* Wild, or undomesticated. The wild Jungle Fowl (*Gallus ferrugineus*) is the *feral* stock of the domestic fowl.

Ferrugin′eous, } (L. *ferrugin′eus*), *n.* or *a.* Rust-red, or the color of
Ferru′ginous, iron-rust. (Medium tint of burnt sienna.) (Plate IV fig 10.)

Fibril′la (L., pl. *fibril′lœ*), *n.* A small fibre.

Fil′ament (L. *filamen′tum*), *n.* A slender or thread-like fibre.

Filament′ous (L. *filamento′sus*), } *a.* Thread-like
Fil′iform (L. *filifor′mis*),

Filopluma′ceous (L. *filopluma′ceus*), *a.* Having the structure of a filoplume

Fil′oplume (L. *filoplu′ma*), *n.* A thread-like feather

Fim′briated (L. *fimbria′tus*), *a.* Fringed

Fissipal′mate (L. *fissipalma′tus*), *a.* With half-webbed feet, the free portion of the toes lobed, as a Grebe's foot.

Fis′siped (L. *fis′sipes*), *a.* Having cleft toes (Opposite of *palmiped*.)

Fissiros′tral (L. *fissiros′tris*), *a.* Having the mouth cleft far back of the base of the bill, as in the Goatsuckers, Swifts, etc., pertaining to the *Fissirostres*

Fissiros′tres (L.), *n.* An obsolete name of an artificial group of birds, with deeply cleft mouths, including the Goatsuckers, Swifts, and other "fissirostral" families.

Flame Red, } (L. *flam′meus, ig′neus*), *n.* A very intense orange-red
Flame Scarlet, color, intermediate between scarlet and saturn-red (Rose carthame + cadmium-orange.) (Pl VII fig 14.)

Flam′mulated (L. *flammula′tus*), *a.* Pervaded with a reddish color

Flanks (L. *hypochon′dria*), *n.* In descriptive Ornithology the most posterior feathers of the sides. (Plate XI.)

Flax-flower Blue, *n.* A delicate light purplish blue color. (French blue + white.) (Plate IX fig 14.)

Flesh-color (L *car'neus, incarna'tus*), *n* A pinkish color, like that observable in the cheeks of a person of fair complexion, carnation (Scarlet-vermilion + white.) (Plate VII fig 18.)

Floc'culent (L *floccula'tus*), *a* In descriptive Ornithology, pertaining to the down of newly hatched or unfledged young birds

Floc'cus (L), *n* The down peculiar to unfledged or newly hatched young birds, in *ptilopædic* birds it covers the general surface and is unconnected with the future plumage, while in *psilopædic* birds it sprouts only from the undeveloped feathers, to the tips of which it is often seen clinging when the latter are considerably grown

Flu'viatile (L *fluvia'tilis*), *a* Pertaining to rivers

Fore'head, ⎰ (L *frons*), *n* Fore part of the top of the head, from the
Front, ⎱ base of the bill to the vertex, or crown (Plate XI.)

Fore'-neck (L *gut'tur*), *n* A rather indefinite and arbitrary term, variously applied, but usually referring to the lower throat and jugulum, though not infrequently to the whole of the space included by the chin, throat, and jugulum. In long-necked birds only does the term become of definite application (See note facing plate XI.)

For'ficate (L *forfica'tus*), *a*. Deeply forked, as the tail of a Kite

Form (L *for'mis*), *n* In a special sense, a sort of non-committal term frequently used by modern writers to designate what is of doubtful rank The term "form" is thus used for what may prove to be a species, or may be only a race, but as to the rank of which the author is in doubt

Fos'sa (L, pl *fos'sæ*), *n* A ditch or groove In descriptive Ornithology, used chiefly in the plural, to denote the depressions in which the nostrils are placed

Fosso'rial, *a* Digging into the earth for a habitation (The Burrowing Owl, *Speotyto cunicularia*, is a fossorial bird.)

Fos'ter-parent, *n* A bird which has reared the young of a parasitic species

Fos'ter-young, *n* The young of a parasitic species which has been reared in the nest of another bird.

Free, *a* Said of a leg with the tibia unconfined within the skin of the body

French Blue, *n* A very rich blue color, deeper than ultramarine (French blue.) (Plate IX fig 6.)

French Gray, *n* A fine light bluish gray color, darker than pearl-gray, lighter and bluer than cinereous (Black + intense blue + smalt-blue + white.) (Plate II fig 17.)

French Green, *n* A very pure rich green color, the typical green (Italian ultramarine + light cadmium.) (Plate X fig 19.)

Fre'num (L), *n* A bridle or marking about the head resembling or recalling a bridle.

Fringe (L. *lo'ma; fim'bria; lacin'ia*), n. A lacerated marginal membrane.
Front'al (L *fronta'tus*), a. Pertaining to the forehead.
Front'let (L *an'tia*), n. The extreme anterior portion of the forehead; usually distinguished by a difference of level (usually more depressed) from the forehead, as in the Woodpeckers. When divided by the base of the culmen (as in the Woodpeckers), the frontal points are called *antiæ*. (See plate XII fig 1.)
Frugiv'orous (L *frugi'vorus*), a. Fruit-eating
Fulig'inous (L *fuligino'sus*), n or a. Sooty brown, or dark smoke-color.
Fulves'cent (L *fulves'cens*), a. Inclining to a fulvous color.
Ful'vous (L *ful'vus*), n. A rather indefinite brownish yellow, or yellowish brown tint, like tanned leather; tawny.
Fur'cate (L *furca'tus*), a. Forked.
Fus'cous (L *fus'cus*), n or a. Dark brown, of a rather indefinite shade
Fu'siform (L *fusifor'mis*), a. Spindle-shaped, or tapering at each end. (Plate XVI. fig. 13.)

G.

Gal'eate (L *galea'tus*), a. Helmeted, or armed or ornamented with a frontal shield, as the Gallinules, Coots, Cassowaries, etc.
Gallina'cea (L), n. A name of the Fowl tribe, or Order *Gallineæ* of some authors.
Gallina'ceous (L *gallina'ceus*), a. Belonging to the Order *Gallinacea* or *Gallineæ*, or that which embraces the domestic fowl and kindred birds Having the characteristics or nature of the *Gallineæ*.
Gall'-stone Yel'low, n. A very strong brownish yellow, somewhat like yellow ochre, but transparent, and much brighter in its paler tints (Aureolin, raw sienna, and cadmium-orange.) (Plate V fig 6.)
Gam'boge Yel'low, n. A pure yellow color, of a lemon tint, less intense and somewhat less pure than the lighter cadmiums, but very transparent The pigment thus called is the concreted juice of the *Hebradendron cambogioides*, a plant which grows in Cambodia (Plate VI fig. 10.)
Gape (L *ric'tus*), n. The opening of the mouth.
Gastræ'um (L), n. The lower parts, collectively
Gen'a (L), n. The cheek, or feathered portion of the lower jaw.
Gen'era, n. Plural of *Genus*.
Gener'ic, a. Pertaining to a *Genus*.
Gen'esis, n. In biological science, the derivation or origin of a form, whether by evolution or direct creation.
Genet'ic, a. Pertaining to *Genesis*.
Ge'nus (pl *gen'era*), n. An assemblage of species which agree in the possession of certain characters distinguishing them from otherwise

allied forms. (In taxonomic value a genus ranks next below a subfamily.)

Gen'ys (L.), *n* (Same as *Gonys*, which see.)

Geograph'ical Race, *n* (See *Race*.)

Geograph'ical Varia'tion, *n.* Modification of form or coloration according to change of locality or country. (The majority of widely distributed species are more or less affected by geographical variation, from varying influences of climate and other surroundings. Many species have evidently sprung from Geographical Races through the extermination of intermediate specimens, or, in the case of remote islands, by long and complete isolation from the parent stock.)

Gera'nium Pink, *n* A lighter tint of geranium red. (Plate VII fig. 19.)

Gera'nium Red (L. *carthami'nus*), *n* The purest possible red color, or a red which combined with yellow will produce a pure orange, and with blue a pure purple. It is less orange in tint than scarlet. (Rose carthame or safflor roth.) (Plate VII fig. 7.)

Gib'bose,
Gib'bous, } (L. *gibbo'sus*), *a* Swollen

Gibbos'ity, *n* A swelling, or rounded protuberance.

Gla'brous (L. *gla'brus*), *a* Smooth

Gla'cial (L. *glacia'lis*), *a.* Pertaining to ice

Glauces'cent (L. *glauces'cens*), *a* Inclining to Glaucous

Glau'cous (L. *glau'cus*), *a* or *n* A whitish blue color, like the "bloom" of a cabbage-leaf. (Black + Antwerp blue + white.) (Plate IX fig. 19.)

Glau'cous Green (L. *glau'co-vir'idis*), *n* (Viridian + white.) (See plate X fig. 17.)

Gnathid'ium (L. pl *gnathid'ia*), *n* The branch or rhamus of the lower jaw, as far as it is covered by the horny sheath. (Chiefly used in the plural.)

Gol'den Yel'low (L. *au'reo-fla'vus; au'rens*), *n* A very intense yellow color, like the paler tints of the pigment called *Jaune d'Or* (that is, golden yellow), which, however, in its deeper tint becomes an intense orange

Go'nys (L.), *n* The keel or lower outline of the maxilla or lower mandible, from the tip to the point where the rhami begin to diverge. (Plate XII fig 6.)

Gorg'et, *n* An ornamented throat-patch, distinguished by color or texture of feathers, as the gorget of a Humming-bird.

Gra'dient (L. *gra'diens*), *a* Walking or running by steps. (Same as *ambulatory*, but preferable to that term.)

Grad'uated,
Grad'uate, } (L. *gradua'tus*), *a* A *graduated tail* has the middle feathers longest, the rest successively shorter; the difference in length not so great, however, as in a *cuneate* tail

Grallato'res,
Grallato'riæ, } (L.), *n* An arbitrary and artificial group of the older classifications, including the wading birds

Grallato'rial, a. Wading, pertaining to the wading birds, or *Grallatores*.
Graminiv'orous (L. *graminiv'orus*), a. Grass-eating. (Geese are *graminivorous*.)
Graniv'orous (L. *gram'vorus*), a. Seed-eating, like certain finches.
Gran'ular (L. *granula'ris*), } a. With a roughened surface, like coarse
Gran'ulate (L. *granula'tus*), } sand-paper.
Grass-Green (L. *viridis, prasi'nus*), n. A deep green color, like growing grass. (Sap-green.) (Plate X fig 4.)
Gray (L. *gris'eus; cæ'sius, cine'reus; ca'nus, leucophæ'us*), n. A color produced by the mixture of black and white. Various shades, dependent on varying relative proportions of the components, are represented on Plate II. figs 2–10.

Greater coverts (L. *tec'trices a'læ ma'jor*),
Greater Wing-coverts (L. *tec'trices seconda'rii*), { n. The most posterior series of wing-coverts, or those which immediately overlay the base of the secondaries, hence, often and very appropriately called *Secondary coverts* (Plate XI.)

Grega'rious (L. *grega'rius*), a. Going in flocks.
Ground-color, n. The prevalent color of the general surface. (Used chiefly in oology.)
Gu'la (L.), n. The throat. (Plate XI.)
Gu'lar (L. *gula'ris*), a. Pertaining to the throat.
Gut'tate (L. *gutta'tus*), { a. Drop shaped or tear-shaped, having
Gut'tiform (L. *guttifor'mis*), { drop-or tear-shaped spots. (Plate XIV. fig 8.)
Gymnopæd'ic, a. Naked at birth. (Synonymous with *ptilopædic*.)
Gymnorhin'al (L. *gymnorhi'nus*) a. Having naked or unfeathered nostrils.

H.

Hab'itat (L. *habita'tus*), n. The region or locality inhabited by a species.
Hab'itus (L.), n. Mode of life.
Hack'le, n. A long lanceolate or falcate feather adorning the neck of the domestic cock. (Used chiefly in the plural, or in combination with *neck*, as *neck-hackles*.)
Hæmatit'ic (L. *hæmati'ticus*) a. Of a blood-red color, crimson.
Hair Brown, n. A clear, somewhat grayish tint of brown, resembling the "brown" hair of human beings, the typical brown color, composed of equal proportions of red and green. (Bistre + raw umber + black + white.) (Plate III. fig. 12.)

Hal'lucal, *a* Pertaining to the hallux, or hind toe

Hal'lux (L.), *n.* In birds possessing four toes, the hinder one is the hallux, no known bird having four toes directed forwards. In some birds, as certain Plovers, the Bustards (*Otididæ*), the *Struthiones*, etc., the hallux or hind toe is wanting. In three-toed birds having two toes in front and one behind, the hallux is usually the one wanting, the hind toe being in reality the fourth (or outer) toe reversed. When the toes are in pairs (two before and two behind), the hallux is usually the inner of the hinder pair, the exception being in the Trogons (*Trogonidæ*). The hallux reaches its best development in the *Passeres*, the *Accipitres*, *Striges*, and *Rallidæ*, but more especially in the first, in which it is usually as strong as if not stronger than the largest of the anterior toes. (Plate XI.)

Ham'ulate (L. *hamulatus*), *a.* Furnished with a small hook

Ham'ulus (L.; pl. *ham'uli*), *n.* A small hook, sometimes applied to the barbules or barbuls of a feather, when hook-shaped.

Hand-quills, *n.* The Primary quills, or primaries

Has'tate (L. *hastatus*), *a.* Shaped like a spear-head (Plate XV. fig. 2.)

Ha'zel (L. *coryllinus*, *avellinus*, *avellaneus*), *n.* An orange-brown color, like the shell of a hazel-nut or filbert; similar to chestnut, but with less red and more yellow. (Vermilion + raw sienna + black.) (Plate IV. fig. 12.)

Heel (L. *suffrago; calcaneus*, *talus*), *n.* The upper posterior extremity of the tarsus (Plate XI.)

He'liotrope Pur'ple, *n.* A grayish purple color (Violet madder-lake + sepia + French blue + white.) (Plate VIII. fig. 18.)

Hel'met (L. *galeatus*), *n.* A naked shield or protuberance on the top or fore part of the head

Hepat'ic (L. *hepaticus*), *a.* Pertaining to the liver, hence, liver-colored

Herodio'nes (L.), *n.* A natural group of altricial waders, embracing the Storks, Wood-Ibises, true Ibises, Spoonbills, Boatbills, and Herons

Herodio'nine, *a.* Pertaining to or partaking of the character of the *Herodiones.*

Her'ring-bone (*markings*), *n.* A series of transverse lines or bars connected along the middle of a feather by a longitudinal stripe or line of the same color (Plate XV. fig. 15.)

Heterodac'tylæ (L.), *n.* The name of a natural group of birds, including only the Trogons

Heteroge'neous, *a.* Of dissimilar nature or miscellaneous character (Opposite of *homogeneous.*)

Hex'agon, *n.* A figure of six sides

Hexag'onal (L. *hexagonalis*), *a.* Having six sides

Hiber'nal (L. *hibernus*), *a.* Pertaining to winter

Hind-neck (L. *cervix*), *n.* (See plate XI.)

Hind-toe (L. *hallux*), *n.* The posterior toe or hallux (which see) (See plate XI).

Hir'sute (L. *hirsutus*), *a.* Hairy, or shaggy, as the foot of a Grouse

Histol′ogy, *n.* Minute anatomy.

Hoar′y (L *al′bens, albes′cens, canes′cens; pruino′sus*), *n* or *a.* Of a frosty gray or silvery hue

Holorhi′nal, *a.* Having the posterior border of the osseous nares rounded. (See *Schizorhinal*)

Homogen′ity, *n.* Structural similarity.

Homoge′neous, *a* Of the same character or nature (Opposite of *heterogeneous.*)

Homolog′ical, } *a* Structurally related or affined. (Opposite to *ana-*
Homol′ogous, } *logical* or *analogous.*)

Homologonat′æ (L), *n* A primary subdivision of the Order *Euripidura*, proposed by Professor A. H Garrod

Homol′ogy, *n* Structural affinity. (Opposite of *analogy*, or superficial resemblance)

Ho′monym, *n* A word which in several senses has different meanings. As *Sylvicola*, Swainson, a genus of birds (now called *Dendroica*) is a homonym of *Sylvicola*, Humphreys, previously applied to a genus of mollusks (Opposite of *synonym*)

Homotyp′ical, *a.* Of the same structural type

Homot′opy, *n.* A particular kind of homology.

Hood′ed (L *cuculla′tus*), *a* Having the head conspicuously different in color from the rest of the plumage

Hor′notine (L *hornoti′nus*), *a* or *n* A young bird in its first year

Hu′meral (L *humera′lis*), *a.* Pertaining to the humerus, or, more generally, to the upper arm.

Hu′merus (L), *n.* The upper arm-bone; or, the whole of the upper arm

Hy′acinth Blue (L *hyacin′thinus*), *n* An exceedingly intense purplish blue color, similar to but richer than smalt blue (Schoenfeld's "violet ultramarine") (Plate IX fig 5)

Hy′brid (L *hybri′dus*), *a* or *n* The progeny resulting from sexual intercourse of distinct species

Hybridiza′tion, *n* Production of hybrids

Hy′bridize, *a* To cross and bear offspring which unite the characters of two species

Hye′mal (L *hiema′lis*), *a* Pertaining to winter.

Hy′oid, *a* Properly, pertaining to the *os hyoides*, or tongue-bone, but frequently applied with reference to the tongue itself.

Hyperbo′rean (L *hyperbo′reus*), *a.* Pertaining to the extreme North

Hyperchrom′atism, *n* State of highly increased brightness or intensity of coloration, or excess of pigment

Hyper′trophy, *n* Unusual development of a part or organ (Opposite of *atrophy*)

Hypochon′drus, } (L , pl *hypochon′dria*). The flanks (Used chiefly
Hypochon′drium, } in the plural) (See plate XI)

Hypochon′driac (L *hypochondria′cus*), *a.* Pertaining to the flanks.

Hypognath′ous, *a* Having the maxilla, or lower mandible, longer than the mandible, as in the Skimmers (*Rhynchops*).

Hypopti'lum (L), *n* An accessory plume, attached to the barrel or stem of ordinary feathers, excepting always the remiges and rectrices (Essentially the same as *after-shaft*.)

Hypora'dii (L ; pl), *n.* Barbs of the hypotilum, or after-shaft

Hyporrha'chis (L), *n.* The after shaft, or stem of the accessory plume, or hypoptilum

Hypoth'esis, *n* A reasonable presumption to account for what is not understood, and hence to be distinguished from *theory*, based upon known facts

Hypothet'ical, *a.* Reasonably presumptive, or probable, though assumed without proof.

I.

Identifica'tion, *n* The determination of the species to which a given specimen belongs

Iden'tify, *v* To determine the systematic name of a specimen

Igno'ble (L *igno'bilis*), *a.* Said of certain Hawks used in falconry Technically, applied to the short-winged Hawks (that is, the Goshawk and Sparrowhawk), to distinguish them from the *noble* Falcons (that is, true Falcons)

Il'iac (L *il'iacus*), *a* Pertaining to the flanks.

Im'bricate, } (L. *imbrica'tus*), *a.* Overlapped, like shingles upon a
Im'bricated, } roof.

Immac'ulate (L. *immacula'tus*), *a.* Entirely free from spots or other markings

Immature', *a* Not adult

Imper'forate (L *imperfora'tus*), *a.* Not pierced through

Incised' (L *inci'sus*), *a.* Cut out, cut away.

Incuba'tion, *n* The act of sitting on eggs in order to hatch them.

Incum'bent (L *incum'bens*), *a* Laid at full length (Said of the hallux, or hind toe, when inserted on a level with the anterior toes)

Indent'ed (L. *indenta'tus*), *a.* Notched along the margin with a different color

In'dian Pur'ple, *n* A very dull purple color, like the pigment of the same name. (Madder-carmine + intense blue + black) (Plate VIII. fig 6)

In'dian Red, *n* A fine rufous-red color, of a slightly more purplish tint than the pigments called Light Red and Venetian Red. Same as brick red (See plate IV fig 11)

In'dian Yellow, *n* A very intense, rich yellow color, much deeper than gamboge, but less pure than cadmium (Plate VII fig 5)

Indig'enous, *a* Native of a country.

In'digo Blue (L *indigo'ticus*) *n* A dark dull blue color, like the indigo of commerce. (Plate IX fig 1)

In'fra- (in composition). Situated under, or beneath. (Opposite of *supra*, — above.)

Infraorb'ital (L. *infraorbita'lis*), *a.* Below the orbit. (Same as *suborbital*, which is more often used.)

Infla'ted (L. *infla'tus*), *a.* Blown out.

Inflex'ed (L. *inflex'us*) *a.* Turned inward.

Infundibu'liform, *a.* Funnel-shaped.

Inguin'al (L. *inguina'lis*), *a.* Pertaining to the groin.

In'ner Toe, *n.* That situated on the inner side of the foot, whether anterior or posterior, but usually the former. (The anterior inner toe is usually the second, but in some zygodactyle forms, as the Trogons, it is the third, the second toe being reversed, thus becoming the inner posterior toe. In a very few — as certain Kingfishers — the second toe is rudimentary or wanting, while in others the first, or hallux, is reversed, and thus becomes the inner anterior toe.) (Plate XI.)

Insectiv'orous (L. *insecti'vorus*), *a.* Feeding upon insects.

Insesso'res (L.), *n.* An obsolete name formerly applied to an artificial group embracing the Passeres and other "perching" birds.

Insesso'rial, *a.* Pertaining to or having the character of perching birds.

Insist'ent, *a.* Said of the hind toe when the greater part of its under surface touches the ground. (Same as incumbent.)

In'stinct, *n.* "A certain power or disposition of mind, by which, independent of all instruction or experience, without deliberation, and without having any end in view, animals are unerringly directed to do spontaneously whatever is necessary for the preservation of the individual or the continuation of the kind."

Integ'ument, *n.* A covering or envelope, usually membraneous, as the skin of animals, the covering of a seed, etc.

In'ter (in composition). Between.

Intermax'illary, *n.* or *a.* The principal bone of the upper jaw, or relating to the same. (Same as *premaxillary*.)

Interorb'ital, *a.* Between the eye-sockets.

Interrham'al, *a.* Between the forks or rhami of the lower jaw.

Interrupt'ed (L. *interrup'tus*), *a.* Discontinued, or broken up.

Interme'diæ, *n.* The middle pair of tail-feathers, or middle rectrices. (Plate XI.)

Interscap'ular (L. *interscapula'ris*), *a.* Between the scapulars.

Interscap'ulars, *n.* The feathers of the interscapulum, or back.

Interscap'ulum (L.), *n.* The region between the scapular tracts, or the back proper. (Plate XI.)

Intertrop'ical, *a.* Between the Tropics, tropical.

Invag'inate (L. *invagina'tus*), *a.* Sheathed.

Inverse', *a.* Inverted, upside down.

I'rian, } *a.* Pertaining to the iris.
Irid'ian, }

Irides'cent (L. *irides'cens*), *a.* With changeable colors, or tints which vary with different inclinations to the light.

I'ris, *n* The (usually) colored circle of the eye surrounding the pupil. (Plate XII fig 11)
Isabel'la-col'or (L *isabelli'nus*), *n* A light grayish cinnamon color, or light buffy brown. (Raw umber + raw sienna + white.) (Plate III fig 23)
Isopo'gonous, *a*. Having the two webs equal in breadth
Isth'mus, *n*. A narrow strip, or neck, connecting two larger areas.

J.

Jug'ular (L *jugula'ris*), *a* Pertaining to the jugulum.
Jug'ulum (L), *n* The lower throat or foreneck, immediately above the breast It is a well-defined area in the Hawks, Vultures, Pigeons, Ducks, and some other groups (Plate XI)

K.

Kid'ney-shaped (L. *renifor'mis*), *a*. Somewhat heart-shaped, but without the point, and broader than long (See *reniform*, plate XIV. fig. 19.)
Knee, *n* Properly the femoro tibial joint, concealed in most birds, but usually the tibio-metatarsal articulation, or *heel*, is so called

L.

Lac'erate (L. *lacera'tus*), } *a*. Jagged, or slashed at the end or along
Lacin'iate (L *lacinia'tus*), } the edge.
Lach'rymal (*bone*), *n* A large bone bounding the orbit anteriorly and above, it is especially well-developed in certain *Falconidæ*.
Lacus'trine (L *lacus'tris*), *a* Lake-inhabiting
Lake Red, *n*. A purplish red color, not so intense as crimson. (Medium tint of madder-carmine) (Plate VII fig 2)
Lamb'doid, *a* L-shaped
Lamellıros'tral (1, *lamelliros'tris*), *a* Having a lamellate bill
Lamelliros'tres (L), *n* A group of birds embracing the *Anatidæ* and Flamingoes, in which the bill is lamellate edged.
Lam'ina, } (L), *n* A thin plate or scale
Lamel'la, }
Lam'inate (L *lamina'tus*), } *a*. Plated, or scaled.
Lam'ellate (L *lamella'tus*), }
Lan'ceolate (L *lanceola'tus*), *a* Lance-shaped, tapering gradually to a point at one end, and more abruptly at the other. (Plate XIV. fig. 12)

Lanu'ginous (L. *lanugino'sus*), *a.* Woolly.

Lat'eral (L. *latera'lis*), *a.* Towards or on the side, pertaining to the side of anything.

Lat'erally, *a.* Sidewise, toward the side.

Lav'ender (L. *lavendula'ceus*), *n.* A very pale purplish color, paler and more delicate than lilac. (Violet + white.) (Plate VII fig 16.)

Lav'ender-Gray (L. *lavendula'ceo-ca'nus*), *n.* (Black + white + smalt-blue.) (Plate II fig 19.)

Lead-col'or (L. *plum'beus*), *n.* (See plumbeous.) (Plate I fig 15.)

Leg, *n.* As generally used, synonymous with *tarsus*, as, "legs and feet," = tarsi and toes.

Lem'on Yel'low (L. *cit'reus; citrin'us*), *n.* A very pure light yellow color, much like gamboge, but purer and richer. (Schoenfeld's "heller cadmium".) (Plate VI fig 11.)

Les'ser Wing-cov'erts (L. *tec'trices mino'res*), *n.* The smaller wing-coverts, forming a more or less well-defined tract immediately anterior to the middle coverts, and thence to the anterior border of the inner wing. (Plate XI.)

Li'lac, } (L. *lilaci'nus, lila'ceus*), *n.* A light purple color, like the
Lila'ceous, } flowers of the lilac. (Purple + white.) (Pl VIII fig 19.)

Li'lac-Gray (L. *lilaci'no-ca'nus*), *n.* (Lamp-black + white + cobalt blue + madder-carmine.) (Plate II fig 18.)

Lim'bate (L. *limba'tus*), *a.* Edged with a different color.

Limico'læ (L.), *n.* The group of shore-birds, a more or less natural group, embracing the Plovers, Sandpipers, Snipe, Curlew, etc.

Limic'oline (L. *limico'lus*), *a.* Shore-inhabiting. Pertaining to, or having the character of, the *Limicolæ*.

Lin'ear (L. *linea'ris*), *a.* Narrow, with straight parallel edges, line-like. (Plate XIV fig 10.)

Lin'eate (L. *linea'tus*), *a.* Marked with lines.

Lin'eolate (L. *lineola'tus*), *a.* Marked with little lines.

Li'ning of the Wing, *n.* The under wing-coverts collectively, especially the lesser and middle. (Plate XIII. fig. 1.)

Lit'toral (L. *litto'ralis, litora'lis*), *a.* Pertaining to the sea shore.

Liv'er Brown (L. *hepa'ticus*), *n.* A dark purplish brown color, like raw liver. (Vermilion + black.) (Plate IV fig 4.)

Lo'bate, } (L. *loba'tus*), *a.* Furnished with membraneous flaps, as the
Lobed, } toes of a Coot (*Fulica*).

Lobe (L. *lo'bus*), *n.* A membraneous flap.

Long-exsert'ed, *a.* Said of tail-feathers when abruptly much longer than the rest.

Longipen'nes (L.), *n.* A group of long-winged swimming birds, formerly embracing the gulls and their allies, and the *Procellaridæ* (petrels, albatrosses, and fulmars), but properly restricted to the *Laridæ, Rhynchopidæ*, and *Stercorariidæ*.

Longipen'nine (L. *longipen'nis*), *a.* Pertaining to the *Longipennes*.

GLOSSARY OF TECHNICAL TERMS. 89

Longiros'tral (L *longiros'tris*), *a* Having a long bill, or pertaining to the artificial and obsolete group *Longirostres*

Longiros'tres (L), *n* An obsolete group of birds, embracing certain long-billed forms

Longitud'inal (L *longitudina'lis*), *a* Running lengthwise, or in the direction of the antero-posterior axis of a body or object.

Lo'ral (L. *lora'lis*), *a* Pertaining to the lores (Plate XII. fig 16)

Lore (L `lo'rum*), *n* The space between the eye and bill in birds (Plate XI)

Low'er Parts (L *gas'træum*), *n* The entire under surface of a bird, from the chin to the crissum, inclusive (See plate XI, and note facing the same)

Low'er Tail-cov'erts (L *tec'trices cau'dæ inferio'res; tec'trices subcauda'les*), *n* The feathers immediately underneath the tail (See *Crissum.*) (Plate XI.)

Lum'bar, *a*. Pertaining to the loins

Lu'minous (L *lumino'sus*), *a* Brilliantly shining, emitting light

Lu'nulate (L *lunula'tus*), *a* Narrowly crescent-shaped (Plate XV fig 5)

Lu'nule (L *lu'nulus*), *n* A small or narrow crescent

Lur'id (L *lur'idus*), *a*. "A color between purple, yellow, and gray;" livid.

Lu'teous (L *lu'teus*), *a* Yellowish, more or less like buff or clay-color

Ly'rate (L *lyra'tus*), *a* Shaped like a lyre, as the tail of the male Blackcock (*Lyrurus tetrix*), or that of the Lyre-bird (*Menura superba*).

M.

Mac'ula (L *ma'cula*), *n* A spot.

Mac'ulate (L *macula'tus*), *a* Spotted.

Mad'der Brown, *n* A very rich reddish brown color, more purplish than burnt sienna (Purple madder + burnt sienna) (Plate IV fig 3)

Magen'ta,
Magen'ta Pur'ple, } *n.* An exceedingly rich reddish purple color, similar to solferino, but darker. (Anilinrosa or rose aniline + aniline violet) (Plate VIII fig 14)

Maize Yel'low, *n* A delicate pale yellow, similar to Naples Yellow, but paler, more creamy than primrose-yellow (Light cadmium + white) (Plate VI fig 21)

Ma'la (L), *n* The side of the lower jaw, behind the horny covering of the mandible.

Mal'achite Green, *n* A light green color, like the mineral called malachite (Italian ultramarine + light cadmium + white) (Plate X fig 6)

Ma'lar (L. *mala'ris*), a. Pertaining to the *mala*. (Plate XII figs. 3, 19.)

Ma'lar A'pex (L. *an'gulus malar'is*), n. The extreme anterior point of the malar region. (Plate XII fig. 3.)

Ma'lar Re'gion (L. *re'gio mala'ris*), n. The side of the lower jaw behind the horny covering of the mandible, usually feathered. In most birds it is a well-defined tract, extending backward from the base of the maxilla, beneath the lores, orbits, and auriculars; and bounded beneath by the chin and throat. (Plate XI.)

Man'dible (L. *mandi'bula*), n. The jaw, when not otherwise indicated, the lower part of the bill is understood. (Plate XI.)

Mandib'ular (L. *mandibula'ris*), a. Pertaining to the mandible.

Man'tle (L. *pal'lium*; *stra'gulum*), n. In certain *Laridæ* and some other birds, *the mantle* is that portion of the upper plumage distinguished from the other parts by a peculiar and uniform color, suggesting, by its position, a mantle thrown over the body. It usually includes simply the back, scapulars, and wings, and the term is perhaps appropriate only when thus restricted. (See plate XI, and note facing the same.)

Mar'bled (L. *marmora'tus*), a. Distinctly varied with irregular markings, or a confused blending of irregular spots, streaks, etc.

Mar'bling, n. Markings which resemble, or suggest, the variegation of marble. In *marbling*, as applied to the plumage of birds, the markings are much more definite and distinct than in clouding, or *nebulation*.

Marine' (L. *mari'nus*), a. Pertaining to the sea.

Marine' Blue, n. A very rich dark blue color. (Winsor & Newton's "intense blue".) (Plate IX fig. 2.)

Mar'gined (L. *margina'tus*), a. Narrowly bordered with a different color.

Maroon' (L. *a'tro-purpu'reus*, *a'tro-coccin'eus*), n. A rich brownish crimson, nearly like the pigment called Purple Madder, claret color. (Madder-carmine + purple madder.) (Plate IV fig. 2.)

Maroon'-Pur'ple, n. See plate VIII fig. 9. (Madder-carmine + purple madder.)

Mars Brown, n. A bright, somewhat yellowish brown color, nearly intermediate between cinnamon and mummy brown. (Sepia + burnt umber + orange-cadmium.) (Plate III fig. 13.)

Mask'ed (L. *persona'tus*, *larva'tus*, *capistra'tus*), a. Having the anterior portion of the head colored differently, in a conspicuous manner, from the rest of the plumage.

Max'illa (L.), n. The jaw, but best restricted to the upper jaw, sometimes called upper mandible. (Plate XI.)

Max'illar, } (L. *maxilla'ris*), a. Pertaining to the maxilla or upper bill.
Max'illary, }

Mauve (L. *mahra'ceus*; *malvi'nus*), n. A light tint of violet. (Aniline violet + white.) (Plate VIII fig. 13.)

Me'dian, } (L. *media'nus*), a. Along the middle line.
Me'dial, }

Mel′anism (L. *melanis′mus*), *n.* A peculiar state of coloration resulting from excess of black or dark pigment. The normal colors of the plumage are replaced by a more or less continuous black or dusky color. The opposite extreme of color from *albinism*, and of frequent occurrence in the family *Falconidæ*.

Melanis′tic (L. *melanis′ticus*),
Melanot′ic, } *a.* Affected with melanism.

Mem′brane (L. *mem′brana*), *n.* A thin, flexible integument or skin, as the webs between the toes of ducks, etc.

Mem′braneous, *a.* Of a soft skinny nature, as the soft skin about the base of the bill of pigeons, the webs between the toes in ducks, etc.

Men′tal (L. *menta′lis*), *a.* Pertaining to the chin, or *mentum*.

Men′tal A′pex (L. *an′gulus menta′lis*), *n.* The extreme anterior point of the chin. (Plate XII fig. 4.)

Men′tum (L.), *n.* The chin, or anterior part of the space between the rhami of the lower jaw.

Me′sial, *a.* Along the middle line. (Same as *medial*.)

Meso- (*in composition*) Middle, median.

Mesorhin′al (L. *mesorhi′nus*), *a.* Situated between the nostrils.

Metacar′pal (L. *metacarpa′lis*), *a.* Pertaining to the hand, or *metacarpus*.

Metacar′pus (L.), *n.* The hand, exclusive of the fingers; the segment of the wing between the carpus and digits.

Metagnath′ous (L. *metagna′thus*), *a.* Cross-billed, with the points of the maxilla and mandible crossing on the right and left.

Metal′lic (L. *metal′licus*), *a.* As applied to colors having a brilliant appearance, like burnished metal.

Metatar′sal, *a.* Pertaining to the *metatarsus*.

Metatar′sus (L.), *n.* That portion of the leg of birds which in descriptive Ornithology is called the *tarsus*, or that portion, usually unfeathered, which extends from the toes to the so-called "knee" (that is, the heel).

Mid′dle Toe, *n.* The middle one of the three anterior toes. It is usually 4-jointed, and longer than the lateral toes. In numerical order it is the third, the hind toe, or hallux, being the first, and the inner toe the second. In zygodactylous birds it corresponds to the outer anterior toe, the fourth toe being reversed. (Plate XI.)

Mid′dle Cov′erts,
Mid′dle Wing-cov′erts,
Me′dian Cov′erts, } (L. *tec′trices a′læ me′diæ, tec′trices a′læ perver′sæ*), *n.* The series of coverts, usually in a single transverse row, situated between the lesser and greater, or secondary coverts. They usually overlap one another in the reverse manner from the other coverts, the inner or upper edge being the one exposed. (Plate XI.)

Migra′tion (L. *migra′tio*), *n.* Periodical change of abode, influenced chiefly by seasonal changes in climate, in which case the migration is regularly *periodical*, the vernal or spring migration being in the northern hemisphere, northward, the autumnal migration southward, but *vice versa* in the southern hemisphere. The migrations of many birds, however, are *irregular* or *erratic*, being prompted by the necessity of finding the requisite food-supply. The Passenger Pigeon (*Ectopistes migratoria*), American Robin (*Merula migratoria*), Cedar-bird (*Ampelis cedrorum*), etc, are migratory in this sense, while the Tanagers, Orioles, and others, which pass the summer only in northern latitudes and the winter entirely within the tropics, are periodical migrants

Mimet'ic (L *mimet'icus*), *a* Imitative, pertaining to or given to mimicry

Mim'esis (L), { *n* Mockery, or imitation of voice, shape, color, etc.
Mim'icry, { The term *protective mimicry* is applied to animals which imitate in color or shape objects by which they are surrounded or species with which they are associated.

Mir'ror (L *spec'ulum*), *n* A name occasionally given to the *speculum* or metallic wing-spot of ducks, etc

Mol'lipilose (L. *mollipilo'sus*), *a* Softly downy.

Monog'amous, *a*. Mating with a single individual of the opposite sex. Applied to species which pair. Those in which the male assists in incubation and rearing the young are *doubly monogamous*

Monog'amy, *n*. The state of pairing, or having a single companion.

Mon'ograph, *n*. A special treatise upon a given subject, as, a *Monograph of the Woodpeckers*, a *Monograph of the Genus Sylvia*, a *Monograph of the Great Auk*, etc

Monomorph'ic, *a* Of essentially the same or similar type of structure (Opposite of *polymorphic*)

Monoto'kous, *a*. Laying a single egg, as the Petrels, Auks, etc. (Same as *uniparous*)

Morpholog'ical, *a* Pertaining to morphology

Morphol'ogy, *n* The science which treats of the laws of form, or the principles of structure Morphology is the basis of *homology*, while analogy is based upon *teleology*

Mouse Gray (L *muri'no gris'eus*, *murinus*), *n* (Lamp-black + white + sepia) (Plate II fig 11)

Moustache' (L *mys'tar*), *n*. In descriptive Ornithology any conspicuous stripe on the side of the head beneath the eye

Mu'cronate (L *mucrona'tus*), *a* Spine-tipped, as the rectrices of the Chimney-swift (*Chætura pelagica*)

Mucron'ulate (L. *mucronula'tus*), *a* Tipped with small points.

Multip'arous, *a*. Producing many eggs.

Mum'my Brown, *n* A bright brown color, nearly intermediate in tint between burnt umber and raw umber The pigment of this name is prepared from ground Egyptian mummies (Mummy, also, sepia + raw umber + burnt sienna) (Plate III fig 10)

Mu'ral (L *mura'lis*), *a*. Pertaining to a wall

Mu'ricate, } (L. *murica'tus*), *a.* Clothed with sharp points, or prickles
Mu'ricated, }
Myr'motherine (L. *myrmotheri'nus*), *a.* Applied to birds which feed upon ants
Myr'tle Green, *n.* A dark bluish green color, like the upper surface of leaves of the myrtle (*Myrtus communis*). (Schoenfeld's "dark zinnober green," or Winsor & Newton's "Prussian green.") (Plate X fig. 2.)

N.

Nape (L. *nu'cha*), *n.* The upper portion of the hind-neck, or cervix
Na'ples Yel'low, *n.* A very pale ochrey yellow, varying in shade from a very pale buff (as in the pigment called French Naples-yellow) to a deep yellowish buff or straw-yellow tint (as in the English pigment) (Plate VI fig. 18.)
Na'ris (L.; pl *nu'res*), *n.* The nostril. The *external nares* open upon some part of the maxilla or upper mandible. In some birds (as the Pelicans, Cormorants, and other *Steganopodes*, and the Toucans, they are basal and more or less obsolete; in others, as the Woodpeckers and members of the Crow family, they are concealed by the antrorse frontal tufts of feathers. The *internal nares* open as longitudinal slits in the posterior portion of the palate
Na'sal (L. *nasa'lis*), *a.* Pertaining to the nostrils
Na'sal oper'culum, *n.* The scale or hardened membrane overhanging the nostril in some birds. (Plate XII fig 9.)
Nas'cent, *a.* Beginning to grow or exist, or in process of development. A *nascent species* is one which is yet connected with the ancestral stock by individuals of intermediate character. Well-known examples may be cited in the *Colaptes auratus* and *C. mexicanus*, which possess very uniform and pronounced characteristics of color, etc., but are connected by specimens of intermediate characters, formerly supposed to be hybrids, but which are now with good reason believed to be merely representatives of the ancestral stock, and tending more or less toward one or the other of the extremes of differentiation represented by the above-named *nascent species*.
Nata'tion, *n.* Act of swimming.
Natato'res (L.), *n.* Swimming birds, as geese, ducks, gulls, etc.
Natato'rial (L. *natato'rius*), *a.* Capable of swimming; pertaining to the act of swimming, or to swimming birds
Navic'ular (L. *navicula'ris*), *a.* Boat-shaped
Nearc'tic (L. *nearc'ticus*), *a.* Pertaining to the northern portion of the New World or Western Hemisphere. The *Nearctic Realm*, or *Region*, is a primary zoo-geographical division of the earth's surface,

made with reference to the natural distribution of animals, and is essentially coincident in area with the North American continent

Neb'ulated (L. *nebulo'sus*), *a.* Clouded, or indistinctly marked with faint, indefinite, and irregular colors

Ne'moral (L *nemora'lis*), *a.* Pertaining to a wood or grove

Neogæ'an, *a.* Pertaining to the Western Hemisphere or New World

Neossol'ogy, *n.* The study of young birds

Neotrop'ical (L *neotropica'lus*), *n* Pertaining to the tropical portions of America, or the New World

Nidifica'tion, *n* Nest-building, or nesting habits

Nile Blue, *n.* A very delicate fine light greenish blue color. (Schoenfeld's "lichtblau") (Plate IX fig 23)

No'menclature, *n* The names of things, according to a recognized principle of naming, or those peculiar to any department of science Various systems of nomenclature have been employed in the naming of animals and plants. Previous to the institution of the *binomial* system by Linnæus (first promulgated as to zoology in 1758), the *polynomial* system, or the use of several terms as the name of a species, was much in vogue That now employed is the *binomial* system of Linnæus, in which usually only two terms are used, the one generic, the other specific, but occasionally modified, according to the requirements of modern science, by the use of a third term after the specific one, for the designation of nascent species, or "subspecies"

Nor'mal (L *norma'lis*), *a* Usual, regular, or in conformity with a particular rule or standard.

Nos'tril (L *na'ris*, pl *na'res*), *n.* The external openings of the organs of respiration

Notæ'um (L), *n* The Latin equivalent for "Upper Parts"

Nu'cha (L), *n* The nape, or upper part of the cervix (Often, but incorrectly, used for the whole cervix)

Nu'chal (L *nucha'lis*), *a* Pertaining to the nape.

Nup'tial or'naments (L *ornamen'ta nuptia'lia*), *n* As distinguished from *nuptial plumes*, any temporary growth from the unfeathered portion of a bird, characteristic of or peculiar to the breeding season. The compressed maxillary process of the American White Pelican (*Pelecanus erythrorhynchos*), and the accessory or supernumerary portions of the bill in many *Alcidæ*, are among the best-known examples

Nup'tial plu'mage (L *ves'tis nuptia'lis*), *n.* A particular plumage, peculiar to the breeding season, characteristic of some birds

Nup'tial plumes (L *plu'mæ nuptia'les*), *n.* Ornamental feathers acquired at the approach of the breeding season, and cast at the close of that period, as the lengthened plumes of many Herons, the crests and filamentous feathers of some Cormorants, etc

GLOSSARY OF TECHNICAL TERMS. 95

O.

Oar'ed, *a* An *oared foot* (L *stegano'pus*) has the hind toe or hallux united on one side with the anterior toes by a web or connecting membrane Hence the name *Steganopodes*, applied to the group including the Pelicans, Cormorants, etc, in which the feet are of this character

Obcord'ate (L *obcorda'tus*), *a.* Shaped like an inverted heart.

Oblique' (L *obli'quus*), *a.* Slanting, crossing, or running, diagonally.

Ob'long (L *oblon'gus*), *a* Longer than broad

Obome'goid (L *obome'goideus*), *a* Obversely omegoid (Plate XV fig 8)

Obo'vate (L *obova'tus*), *a* Inversely ovate (Plate XIV fig 6.)

Obscure' (L. *obscu'rus*), *a* Dusky, or without distinct definition, little known It is sometimes improperly used in the same sense as *obsolete*, but the two terms are quite distinct in meaning, an *obscure* or *obscured* marking is one which is rendered indefinite by a suffusion with the surrounding color; an *obsolete* marking is one rendered indistinct by lack of intensity or depth of color

Ob'solete (L *obsole'tus*), *a* As applied to words or writings, disused or neglected In the natural-history sense, indistinct, rudimental, faded. An *obsolete* spot, or bar, is one which, while representing a well-developed marking on another individual of the same species, or on another species with which the one being described is compared, is nearly or quite wanting from encroachment of the adjacent color, or lack of intensity of color in itself Hence, *obsolete*, as used in this sense, is quite distinct from *obscure*, often improperly treated as synonymous; an *obscure* spot or other marking being one which lacks distinct definition through suffusion of its own color with that adjacent

Obtuse' (L *obtu'sus*), *a* Blunt (Opposed to *acute*)

Occip'ital (L *occipita'lis*), *a* Pertaining to the hind-head, or occiput

Oc'ciput (L), *n* The back part of the head, bounded below by the nape, anteriorly by the vertex (Plate XI)

Oc'ellate (L *ocella'tus*), *a.* Marked with *orelli*, or eye-spots (Plate XIV. fig 3)

Ocel'lus (L , pl *ocel'li*), *n* A distinct, rounded, usually brightly colored spot, more or less resembling the "eyes," or *ocelli*, of a Peacock's train

Ochra'ceous, ⎫ (L *ochra'ceus*), *a.* Of the color of certain ochre pigments;
Och'reous, ⎬ a brownish orange color, or intense buff (Light
Och'rey, ⎭ ochre, No. 2, of Schoenfeld) (Plate V. fig 7)

Ochra'ceous-Buff (L *ochra'ceo-lu'teus*), *n.* (Yellow ochre + burnt sienna + white) (Plate V fig 10)

Ochra'ceous-Ru'fous (L *ochra'ceo-ru'fus*), n.(Yellow ochre + burnt sienna + light red) (Plate V fig 5)

O'chre-Yel'low (L *ochra'ceo flavus*), n The color of the pigment called yellow ochre. (Plate V fig 9)

Oc'reate (L *ocrea'tus*), a. Booted, or having the anterior covering of the tarsus undivided for the greater part of its length

Odontor'nithes (L.), n. The name of an extinct order or primary group of birds, comprising forms which possessed teeth in sockets, and thus, as well as in other features, more nearly approaching the reptiles in their structure than any living forms

Oil Green (L *oleagineus*), n A dull light yellowish green (Schoenfeld's yellow green zinnober) (Plate X fig 21)

Olfac'tory, n Pertaining to the sense of smell

Oligoto'kous, a. Producing few eggs

Oliva'ceous, Olive, (L. *oliva'ceus; olivi'nus*), n A greenish brown color, like that of olives (Sepia + light zinnober green) (Plate III fig 9)

Ol'ive-Buff (L. *oliva'ceo-lu'teus*), n (Yellow ochre + cobalt blue + white) (Plate V. fig 12)

Ol'ive-Gray (L *oliva'cco-ca'nus*), n. (Black + white + light cadmium) (Plate II fig 14)

Ol'ive-Green (L *oliva'ceo-vi'ridis*), n A peculiar color, common in birds (especially the Warblers, and hence sometimes called "warbler-green"), produced by the mixture of yellow and gray, resulting in a tint somewhat between olive and dull yellowish green. (Light zinnober green + raw umber) (Plate X. fig 18)

Ol'ive-Yel'low (L *oliva'ceo-fla'vus*), n. (Light cadmium + black + white) (Plate VI fig 16)

Omniv'orous (L *omniv'orus*), a. Feeding upon anything eatable; eating indiscriminately

Ome'goid (L *ome'goideus*), a. Resembling in form the Greek capital letter *Ome'ga*, Ω

Oolog'ical, a. Pertaining to oology.

Oöl'ogy, n The science of birds' eggs.

Opales'cence, n. A reflection of pearly tints from a pale or milky ground-color

Opales'cent (L. *opales'cens, margarita'ceus*[1]), a. Reflecting changeable tints from a pale or milky ground color

Opaque' (L *opa'cus*), a In descriptive Ornithology, the opposite of metallic, or brilliant. Dull, or without gloss.

Oper'culum (L), n. A lid, or cover, such as the scale overhanging the nostrils (*operculum naris*) of many birds

Ophthal'mic (L *ophthal'micus*), a. Pertaining to the eye.

Op'tic, a. Pertaining to the sight

---

[1] Properly, this term means pearly, but as used in descriptions the terms are essentially synonymous.

O'ral (L. *ora'lis*), *a.* Pertaining to the mouth
Or'ange (L *auran'tius*), *n* A deep reddish yellow, like the rind of an orange. (Winsor & Newton's cadmium-yellow) (Plate VI fig 3)
Or'ange-Buff (L. *auran'tio-lu'teus*), *n* (Cadmium-orange + white) (Plate VI fig 22)
Or'ange-Chrome, ⎫
Or'ange-Red, ⎬ (L *auran'tio-ru'brum*, *flam'meus*, *ig'neus*), *n.* A fine bright light red color, verging somewhat to orange, like the pigment called orange-chrome. (Plate VII fig 13)
Or'ange-Ochra'ceous (L. *auran'tio-ochra'ceus*), *n.* (Cadmium-orange + yellow ochre + burnt sienna) (Plate V fig 3)
Or'ange-Ru'fous (L *auran'tio-ru'fus*), *n* (Neutral orange, or cadmium-orange + light red) (Plate IV fig 13)
Or'ange-Vermil'ion (L. *auran'tio-cinnabari'nus*), *n* See plate VII fig. 12 (Scarlet-vermilion + orange-cadmium)
Or'ange-Yellow (L *auran'tio-fla'vus*), *n* A color intermediate between orange and yellow
Orbic'ular (L *orbicula'ris*), *a.* Circular
Or'bit (L. *or'bitus*), *n* The region immediately around the eye.
Or'bital Ring, *n* A ring or circle of color immediately surrounding the eye (Plate XII fig. 12.)
Or'der (L *or'do*, pl *ordines*), *n.* In natural history, a group of families possessing in common peculiar characteristics
Or'dinal, *a.* Pertaining to an order
Ornith'ic, *a.* Pertaining to birds.
Ornithol'ogy, *n* The science of birds.
Ornithot'omy, *n* The anatomy of birds.
Or'piment Or'ange, *n* A deep dull orange color, much less pure than cadmium (Cadmium-orange + burnt sienna) (Plate VI fig 1)
Os'cinine, *a.* Pertaining to the *Oscines*: musical, or capable of singing.
Os'cines (L), *n* The name of a natural group of singing passerine birds, comprising the singing-birds *par excellence*, characterized by a highly specialized vocal apparatus. (Same as *Polymyodæ*)
Os'seous, *a* Bony.
Os'sified, *a.* Become bony.
Osteolog'ical, *a.* Pertaining to osteology
Osteol'ogy, *n* The science of bones, description of the bones or the bony structure of animals, also, the osseous system
Out'er Web (L *pogo'nium exte'rius; pogo'nium externum*), *n* The outer web of a feather is that farthest from the central line of the body; in wing-feathers it is that farthest from the base of the wing, or toward the outer edge of the wing.
Out'er Toe, *n* See plate XI
O'val (L *ova'lis*), *a* Shaped like the longitudinal outline of an egg which has both ends of equal or of similar contour. (Plate XIV. fig 5; plate XVI. fig. 11)

O′vate, (L *ova′tus*), *a.* Shaped like an egg which has one end more
O′void, pointed than the other (Plate XIV fig 7, plate XVI
Ovoid′al, fig 1)
Ova′rium (L , pl *ova′ria*),
O′vary (pl *ova′ries*), *n.* The organ in which eggs are developed
O′viduct (L), *n.* The tube through which the egg passes from the ovary
Ovip′arous, *a.* Producing eggs in which the young develop after exclusion from the body
Oviposi′tion, *n.* Act of laying eggs

P.

Palæarc′tic (L *palæarc′ticus*), *a.* Pertaining to the northern portion of the Eastern Hemisphere, or Old World
Palæogæ′an, *a.* Pertaining to the Eastern Hemisphere, or Old World
Palæornithol′ogy, *n.* The science of fossil birds
Pal′ama (L), *n.* The web or membrane between the toes of certain birds.
Pal′atal,
Pal′atine, *a.* Pertaining to the palate
Pal′ate (L *pal′atum*), *n.* The roof of the mouth.
Pal′ea (L), *n.* A dewlap, or fleshy pendulous skin on the throat or chin, as in a Turkey or domestic fowl.
Pal′lium (L), *n.* A mantle.
Pal′mate,
Pal′mated, (L *palma′tus*), *a.* Having the three anterior toes full-webbed. (Compare *Semipalmate* and *Totipalmate*.)
Pal′miped,
Pal′pebra (L), *n.* The eyelid
Pal′pebral (L *palpebro′sus*), *a.* Pertaining to the eyelids
Pal′pebrate (L *palpebra′tus*), *a.* Having eyelids.
Palu′dicole (L *paludi′colus*), *a.* Marsh-inhabiting.
Pal′udine (L *paludi′nus*),
Palus′trine (L *palus′tris*), *a.* Pertaining to a marsh or swamp.
Pan′durate (L *pandura′tus*), *a.* Fiddle-shaped (Plate XIV.
Pandu′riform (L *panduriformis*), fig 18)
Pan′sy Pur′ple, *n.* An exceedingly rich and very intense deep purple color, like that of some varieties of the pansy (*Viola tricolor*) (Intense blue + madder carmine + rose tyrien) (Plate VIII fig 5)
Pap′illa (L pl *papil′læ*), *n.* A small nipple-like elevation
Pap′illose (L, *papillo′sus*),
Pap′illate (L *papilla′tus*), *a.* Having papillæ
Pap′ula (L.; pl *pap′ulæ*), *n.* A pimple, or pimple-like elevation

Pap'ulous, \
Pap'ulose, } (L. *papulo'sus*), *a.* Pertaining to or having pimples

Paragnath'ous, *a.* Having both mandibles of equal length, the tips meeting.

Par'asite, *n.* In Oology, a species which constructs no nest and performs none of the duties of incubation or rearing of the young, but imposes on other birds for this purpose. A parasitic bird is also a species which obtains its food by systematically robbing other species; as the Parasitic Jaeger (*Stercorarius parasiticus*), Bald Eagle (*Haliæetus leucocephalus*)

Parasit'ic (L. *parasit'icus*), *a.* Depositing the eggs in the nests of other birds, to which are left the duties of incubation and care of the young The European Cuckoo (*Cuculus canorus*) and the common Cow Blackbird (*Molothrus ater*) are well-known examples

Par'is Blue, *n.* A rich blue color, nearly intermediate between Berlin blue and Antwerp blue (Schoenfeld's Paris blue) (Plate IX fig. 7)

Par'is Green, *n.* The finest and most intense of all green pigments (Paris green) (Plate X fig 13)

Parot'ic, } (L. *parot'icus*), *a.* Pertaining to the region immediately \
Parot'id, } beneath the ear

Par'rot Green (L. *psitta'ceus, psittacin'us*), *n.* A rich, somewhat yellowish green color, like the plumage of many species of Parrots (Schoenfeld's light green zinnober) (Plate X fig 7)

Pas'seres (L), *n.* A group of birds including the most highly developed forms, such as the Thrushes, Warblers, the Sparrow tribe, Crow family, etc , but not the Swifts, Humming birds, Kingfishers, Woodpeckers, etc , which belong to entirely distinct orders

Pas'serine (L *passeri'nus*), *a.* Pertaining to or having the characters of the Passeres

Pea Green, *n.* A pale dull green color, like the color of green pea-pods (Sap green + white) (Plate X. fig 9)

Peach-blossom Pink, *n.* A delicate light pink color, of a more fleshy tint than rose pink (Schoenfeld's pink madder) (Plate VII fig 21)

Pearl Blue, *n.* A very pale purplish blue color. (White + French blue) (Plate IX fig 17)

Pearl Gray (L. *margarita'ceus; margarita'cco-ca'nus*), *n.* A very pale, delicate blue-gray color, like the mantle of certain gulls. (White + intense blue) (Plate II fig. 20)

Pec'tinate, } (L *pectina'tus*), *a.* Having tooth-like projections like the \
Pec'tinated, } teeth of a comb, as the toes of the grouse.

Pectina'tion, *n.* Comb-like toothing

Pec'toral (L *pectora'lis*), *a.* Pertaining to the breast

Pec'tus (L), *n.* The breast.

Pe'des (L ; pl of *pes*), *n.* The feet, which in birds includes the leg below the tibia

Pelag'ic (L *pela'gicus*), *a.* Frequenting the high seas.

Pelas'gic (L *pelas'gicus*), *a*   Wandering.
Pel'ma (L), *n*   The under surface of the foot
Penicil'late (L *penicilla'tus*), *a*   Brush-tipped or pencil-like.
Pen'na (L), *n*   A perfect feather
Penna'ceous (L *penna'ceus*), *a*   Pertaining to a perfect feather, or having the character of the same
Per'forate (L *perfora'tus*), *a*   Pierced through, said of nostrils which communicate with one another by reason of the absence of a septum, as in the American Vultures (*Cathartidæ*).
Peristeromorph'æ (L), *n*   The Huxleyan name, meaning "dove-formed," of the order Columbæ
Per'vious, *a*   Open, used synonymously with perforate, as applied to the nostrils
Pe'trous, *a*   Stony; hard like stone.
Phal'anx (L , pl *phalan'ges*), *n.*   In birds, a joint (not hinge, or articulation) or segment of the toes.
Phase, *n*   Used more especially in the case of dichromatic species, as the *melanistic phase*, the *rufescent phase*, etc
Phlox Pur'ple, *n*   A very fine, medium, or rather light reddish purple, like the color of some varieties of Phlox. (Violet madder + rose tyrien) (Plate VIII fig 11)
Phys'ical, *a.*   Pertaining to the bodily organization
Physiog'nomy, *n*   The general appearance   Properly, the countenance, with respect to the temper of the mind
Physiol'ogy, *n.*   The science of bodily functions
Pi'ci (L), *n*   The name of a natural group, or Order, of zygodactyle birds, comprising the Woodpeckers and Wrynecks.
Pi'cine (L. *pici'nus*), *a*   Pertaining to the Woodpecker tribe, woodpecker-like
Pictu'ra (L., pl *pictu'ræ*), *n.*   Pattern of coloration of a particular part, or a particular feather
Pig'ment, *n*   Coloring-matter
Pil'eate,
Pil'eated,   (L. *pilea'tus*), *a*   Capped, or with the whole pileum crested. Differing from *crested*, in that the latter is used to designate an elongation of the feathers on a *particular part* of the pileum, as a *frontal, vertical,* or *occipital* crest
Pil'eum (L *pi'leus*), *n.*   The cap, or whole top of head from bill to nape, and therefore including the forehead, vertex (or crown), and occiput (Plate XII)
Pil'ose (L. *pilo'sus*), *a.*   Slightly hairy
Pink (L *caryophylla'ceus*), *n.*   A dilute rose-red color. (See Rose Pink, and Peach-blossom Pink)
Pink'ish Buff (L *caryophylla'ceo-lu'teus*), *n*   (Yellow ochre + light red + white) (Plate V fig 14)
Pink'ish Vina'ceous (L. *caryophylla'ceo-vina'ceus*), *n*   (Winsor & Newton's Indian red + white.) (Plate IV. fig. 18.)

GLOSSARY OF TECHNICAL TERMS. 101

Pin'nate, ⎫ (L. *pinna'tus*), *a.* Having wing-like tufts of elongated
Pin'nated, ⎭ feathers on the side of the neck.

Pin'niform (L. *pinnifor'mis*), *a.* Fin-like, as a Penguin's wing

Pin'tailed, *a.* Having the central tail-feathers elongated and narrowly acuminate, as in the male Pin-tail Duck (*Dafila acuta*).

Pisciv'orous (L. *pisci'vorus*), *a.* Feeding upon fish

Pla'ga (L.), *n.* A stripe

Plan'ta (L.), *n.* The posterior face of the tarsus

Plan'tar, *a.* Pertaining to the planta

Plan'tigrade, *a.* Walking on the back of the tarsus.

Plas'tic, *a.* Capable of being moulded, easily modified.

Plum Pur'ple, *n.* A rich dark violet-purple (Madder carmine + intense blue) (Plate VIII. fig 4.)

Plu'ma (L.), *n.* A feather.

Plu'mage (L. *indumen'tum*), *n.* The feathering in general

Plum'beous (L. *plum'beus*), *n.* A deep bluish gray color, like tarnished lead, lead-color (Lamp-black + intense blue + white) (Plate II fig 15)

Plum'iped (L. *plum'ipes*), *a.* Having the feet feathered

Plu'mose (L *plumo'sus*), *a.* Feathered.

Plu'mula (L.), *n.* A down-feather

Plumula'ceous (L *plumula'ceus*), *a.* Downy; bearing down

Po'dium (L.), *n.* The foot

Podothe'ca (L.), *n.* The whole envelope of the legs and feet.

Pogo'nium (L., pl *pogo'nia*), *n.* The web of a feather

Pol'lex (L.), *n.* The thumb. In birds, the joint (homologous with the index-finger of man) which bears the alula, or bastard-wing

Polyg'amous, *a.* Mating with many females, as the domestic cock

Polymorph'ic, *a.* Many-formed, containing or consisting of many forms, or different types In Ornithology, a species is 'polymorphic" when it presents several distinct phases of coloration in the same locality or within a restricted geographical area Thus, some of the hawks (e g *Buteo swainsoni*) are polymorphic in this sense

Polymyo'dæ, *n.* The name of a natural group of passerine birds, characterized by highly specialized vocal organs (Synonymous with *Oscines*)

Polyno'mial, *a* or *n.* Consisting of several words, as the polynomial nomenclature, by which a species was designated by a descriptive phrase This system of nomenclature preceded the establishment of the *binomial* system, established by Linnæus A name consisting of several words

Polyto'kous, *a.* Producing many eggs, or young. (Synonymous with multiparous)

Pomegran'ate Pur'ple (L *puni'ceo-purpu'reus*, *puni'ceus; phœni'ceus*), *n* A dull reddish-purple color, like the pulp of some varieties of the pomegranate (*Punica granatum*). (Madder carmine + violet madder) (Plate VIII. fig 12)

Pop'py Red, *n.* A very intense red color, intermediate between vermilion and carmine. (Bourgeois's "laque ponceau") (Plate VII. fig 9.)

Poste'rior (Upper or Lower) Parts, *n* The hinder half of a bird, above or below.

Poste'rior Toe, *n* In most birds, the hallux, or hind toe. In some, however, one of the "anterior" toes is directed backward, and also becomes a posterior, or hind, toe

Postoc'ular (L *postocul'aris*),
Postor'bital (L *postor'bitalis*), } *a.* Back of, or posterior to, the eye The former is most used. (Plate XII fig 17.)

Pow'der-down Feath'ers, *n* Peculiar, imperfect feathers, which grow in matted patches, usually on the interspaces between the true feather-tracts, characterized by a greasy texture and scurfy exfoliation. They are particularly characteristic of the Heron tribe (*Ardeidæ*), but are found in other groups also

Præco'ces (L.), *n* A more or less artificial group of birds, whose young run about and feed themselves immediately after emerging from the egg. The group is composed of the orders *Gallineæ, Limicolæ, Alectorides, Anseres, Pygopodes,* and *Struthiones.*

Præco'cial, *a* Having the nature of, or pertaining to, the *Præcoces.*

Pressiros'tral, *a* Pertaining to the *Pressirostres*

Pressiros'tres (L.), *n* The systematic name of a Cuvierian artificial group of grallatorial birds with hard and compressed bill, comprising the Plovers, Cranes, etc.

Pri'mary (L *rem'iges prima'riæ*), *n* Any one of the quill-feathers of the "hand-wing," usually nine to eleven in number. Used chiefly in the plural, as distinguished from the *secondaries*, or those remiges which grow upon the forearm (Plate XI.)

Pri'mary Cov'erts (L *tec'trices prima'riæ*), *n* The series of stiff feathers, usually corresponding with the primaries in their graduation, which overlie the basal portion of the latter (Plate XI.)

Prim'rose Yel'low (L *primula'ceo-fla'vus*), *n* A very delicate pale yellow, of a more creamy tint than sulphur-yellow (Pale cadmium + white) (Plate VI fig. 13.)

Proce'res,
Proce'ri, } (L.), *n.* A name given by Illiger to the *Struthiones.*

Protrac'tile,
Protru'sile, } *a.* Capable of being thrust forward or elongated, as the tongue of a Woodpecker or a Humming-bird

Prout's Brown *n.* A medium brown color, or the typical brown, composed of equal proportions of green and red (Madder carmine + vermilion + pale cadmium + Italian ultramarine, or, Winsor & Newton's "Prout's brown.") (Plate III fig. 11.)

Prune Pur'ple, *n* A dark reddish-purple color, darker and duller than dahlia purple (Purple madder + violet madder) (Plate VIII fig 1.)

Prus′sian Blue, *n.* A very intense and rich blue color, darker and more greenish than ultramarine and cobalt. Similar to, but less pure than, Antwerp blue, and not a reliable color.

Psilopæ′des (L), *n* A more or less artificial group of birds born weak and helpless, and further distinguished by a scant growth of down affixed to the undeveloped pterylæ, or future feathers, to which it is temporarily attached The *Passeres* and most of the *Picariæ* belong to this group (Synonymous with *Gymnopædes*)

Psilopæ′dic, *a* Pertaining to, or having the nature of, the Psilopædes.

Psitta′ci (L), *n* A very natural group of birds, comprising the Parrot-tribe only.

Psittacomorph′æ (L), *n* The Huxleyan name, meaning "Parrot formed," for the order *Psittaci*

Ptery′la (L , pl *pery′læ*), *n* An area or tract of the skin on which feathers grow A "feather tract"

Pterylog′raphy, *n* A description of the plumage, with reference to the distribution of the feather-tracts or pterylæ

Pterylo′sis (L), *n* The plumage, considered with reference to its distribution on the skin

Ptilopæ′des (L), *n.* A more or less artificial group of birds, instituted by Professor Sundevall, including those which at birth are covered with down (Synonymous with *Dasypœdes*)

Ptilopæ′dic, *a* Pertaining to or having the character of *Ptilopœdes* Clothed at birth with down, like the chick of the domestic fowl, a duckling, or a gosling

Ptilo′sis (L), *n.* Plumage

Pul′lus (L), *n.* A chick. Applied to the downy young of Ptilopædic or Præcocial birds

Punc′tate (L *puncta′tus*), *a* Dotted (Plate XV fig 12)

Pu′pil (L *pupil′la*), *n.* The central black (or dark blue) spot or disk of the eye, enclosed within the iris (Plate XII fig 10)

Pur′ple (L. *purpu′reus*), *n* A color intermediate between red and blue, or produced by the combination of these colors

Pygopo′des (L), *n.* A group of swimming birds, containing the families *Podicipididæ*, *Colymbidæ*, and *Alcidæ*, distinguished by the extreme posterior position of the legs.

Pygop′odous, *a* Pertaining to or having the character of the *Pygopodes*

Pyr′iform (L *pyrifor′mis*), *a.* Pear-shaped (Plate XIV fig 17 ; plate XVI fig 7)

Q.

Quadran′gular (L *quadrangula′ris*), *a.* Four-angled, or square.

Quad′rate (L *quadra′tus*), *a* Square (Plate XIV fig 2)

Quill, *n* As generally used, one of the primary remiges ; and perhaps best so restricted.

Qui′nary, *a.* Consisting of, or arranged by, fives. The *quinary system* of classification, formerly much in vogue, presumed five types for each natural group (that is, five species to a genus, five genera to a family, etc.).

Quin′cunx (L.), *n.* A set of five, arranged thus ∴·

R.

Race, *n.* A nascent species, or a "form," which on account of the existence of intermediate specimens cannot be considered a species, no matter how great a degree of differentiation may have been reached. Races are distinguished as "Geographical" and "Local," according as to whether they occupy extensive or limited areas of country. Geographical races are usually correlative with definite geographical areas, being, in fact, the expression of geographical variation.

Ra′dial, *a.* Pertaining to the radius.

Ra′dii Accesso′rii (L), *n.* The barbs of a supplementary feather, or aftershaft.

Ra′dii (L), *n.* The barbs of a perfect feather.

Radio′li (L), *n.* The barbs of the Radii, or barbules.

Radio′li Accesso′rii (L), *n.* The barbules of a supplementary plume, or aftershaft.

Ra′dius (L), *n.* The outer bone of the forearm.

Ra′mus (L; pl *ra′mi*), *n.* A branch or fork, as the ramus of the lower mandible (that is, *mandibular ramus*). (Plate XII fig 5.)

Rapto′res, *n.* An artificial group of birds, including the so-called Birds of Prey.

Rapto′rial, *a.* Pertaining to the Birds of Prey, or having the characteristics of the Raptores.

Raso′res (L), *n.* The name of the *Gallinaceæ* in some of the older classifications.

Raso′rial, *a.* Pertaining to the *Rasores*, or scratching birds.

Rati′tæ (L), *n.* A group of birds, more or less artificial, including those with a flat or unkeeled sternum, and comprising the orders *Struthiones* and *Apteryges*, all other existing birds being included in the Carinatæ, which have a keeled sternum.

Rau′cous (L *rau′cus*), *a.* Hoarse-voiced.

Raw Sien′na, *n.* A bright yellowish brown, like the pigment of the same name. (Plate V fig. 2.)

Raw Um′ber, *n.* A light, rather yellowish brown, similar to the pigment of the same name. (Plate III fig 14.)

Rec′trix (L, pl. *rec′trices*), *n.* Any one of the tail-feathers. (Used chiefly in the plural.) (Plate XI.)

GLOSSARY OF TECHNICAL TERMS. 105

Recur′ved (L. *recurv′us*), a. Curved upward.
Reflect′ed (L *reflec′tus*), a. Turned backward.
Reflec′tion (L *reflec′tio*), n Change of color with different inclination to the light.
Refract′ed (L. *refrac′tus*), a Abruptly bent, as if broken
Re′gion (L. *re′gio*), n. Any portion of the body localized, as the *anal region* (*regio analis*), *dorsal region* (*regio dorsalis*), etc
Re′mex (L ; pl *rem′iges*), n. Any one of the longer wing-feathers Used chiefly in the plural. The remiges are of two kinds, namely, the *primary remiges*, or quills of the hand-wing, and the *secondary remiges*, or quills of the forearm.
Ren′iform (L. *reniform′is*), a Kidney-shaped (Plate XIV fig. 19)
Rep′licate, } (L *replica′tus*), a. Folded over so as to form a groove or channel
Rep′licated, }
Retic′ulate, } (L *reticula′tus*), a Marked with cross-lines like the meshes of a net
Retic′ulated, }
Reticula′tion, n Net-work
Retrac′tile, a. Susceptible of being drawn back and driven forward, as a cat's claw
Retrorse′ (L. *retror′sus*), a Directed backward.
Rhach′is (L , pl. *rhach′ides*), n The shaft of a feather, exclusive of the hollow basal portion, or "barrel"
Rhi′nal (L *rhina′lis*), a. Pertaining to the nose.
Rhomb′oid (L. *rhomboid′eus*), a. Lozenge-shaped (Plate XIV fig. 1)
Ric′tal, n Pertaining to the rictus (Plate XII fig 18)
Ric′tus (L.), n. The gape; sometimes restricted to the corner of the mouth, or *angulus oris* (Plate XII fig. 2)
Rosa′ceous,
Rose Pink,
 { (L. *rosa′ceus*, *pal′lide-ro′seus*; *caryophylla′ceus*), n A very pure purplish-pink color, like some varieties of roses. (Rose carthame + rose tyrien + white) (Plate VII. fig 20)
Rose Pur′ple (L. *rosa′ceo-purpu′reus*), n A light rosy purple hue, like the petals of some roses (Madder carmine + violet ultramarine + white) (Plate VIII fig 20)
Rose Red (L *ro′seus*; *rosa′ceo-ruber*), n The purest possible purplish red color (Rose carthame + rose tyrien) (Plate VII fig 5)
Ros′trum (L), n The beak
Round′ed (L. *rotunda′tus*), a. A *rounded tail* has the central pair of feathers longest, the remainder successively a little shorter A *rounded wing* is one in which the first primary is short, the longest quill being the third, fourth, or fifth, or one nearly midway between the first and last
Roy′al Pur′ple (L. *ianthin′us*), n A very rich intense violet color, verging toward blue (Aniline violet + violet-ultramarine) (Plate VIII fig 7)
Ru′diment, n. A beginning

Rudiment'ary, *a.* Imperfectly developed, as if only begun

Rufes'cent (L *rufes'cens*), *a* Inclining to a rufous color

Ruff, *n* A collar of elongated or otherwise modified feathers round or on the neck

Ru'fous (L. *ru'fus*), *n* A brownish red color, like the pigments called Venetian Red, Light Red, Indian Red, Red Chalk, etc, which represent various shades of rufous. The typical shade is light red (Plate IV. fig 7)

Ru'ga (L), *n* A ridge or wrinkle.

Ru'gose (L. *rugo'sus*), *a* Wrinkled.

Rump (L *uropy'gium*), *n* That portion of the upper surface of the body lying between the interscapulars and upper tail-coverts. (Plate XI)

Rupi'coline (L. *rupi'colus*), *n* Rock-inhabiting.

Rus'set (L *russa'tus*), *n.* A bright tawny-brown color, with a tinge of rusty (Burnt sienna + cadmium orange + raw umber) (Plate III. fig 16)

S.

Saf'fron Yel'low (L *cro'ceus*), *n* A peculiar shade of yellow, like that produced from the infusion of flowers of the saffron (*Crocus sativus*). (Plate VI fig. 4)

Sage Green, *n.* A dull grayish-green color, like leaves of the garden sage (Green oxide of chromium + black + white) (Plate X. fig 15)

Sag'ittate (L *sagitta'tus*), *a* Shaped like an arrow-head (Plate XV. fig 1)

Sali'va (L), *n* Spittle.

Sal'ivary Glands, *n* The organs which secrete the saliva, or spittle.

Sal'mon-Buff (L *salmona'ceo-lu'teus*), *n* (Light red + cadmium orange + white.) (Plate IV. fig 19)

Sal'mon-Col'or (L. *salmona'ceus*), *n.* A color intermediate between flesh-color and orange, like the flesh of the salmon (Saturn red or orange-chrome + white) (Plate VII fig 17)

Sal'tatory, *a* Progressing by leaps, hopping (Opposite to *ambulatory,* or *gradient*)

Sanguina'ceous (L *sanguina'ceus, sanguin'eus*), *n* or *a.* Blood-red (Same as crimson) (Plate VII fig 3)

Sat'urn Red (L. *minia'tus*), *n* A very fine orange-red color, the same as red lead. (Red lead or saturn red.) (Plate VII fig 16)

Saurop'sida (L), *n* A primary group of vertebrate animals comprising birds and reptiles

Sauru'ræ (L), *n* The name of an extinct primary group or order of birds, including the fossil *Archæopteryx.*

GLOSSARY OF TECHNICAL TERMS 107

Saxic'oline (L *saxi'colus*), *a.* Stone-inhabiting; pertaining to, or having the characteristics of, the Stone Chats (*Saxicola*)
Scab'rous, *a* Scabby, scurfy, scaly.
Scal'loped (L *crena'tus*), *a* Cut along the edge, or border, into segments of a circle. (Plate XV fig 21)
Scan'dent (L *scan'dens*), *a* Climbing
Scanso'rial, *a* Capable of climbing, as a Woodpecker. Pertaining to the obsolete group *Scansores*
Scap'ula (L), *n* The shoulder-blade
Scap'ular (L *scapula'ris*), *a* Pertaining to the scapula.
Scap'ular Re'gion (L *re'gio scapula'ris*), *n* The usually well-defined longitudinal area of feathers overlying the shoulder-blade. They lie along each side of the back (whence the feathers of the latter region are frequently called *interscapulars*)
Scap'ulars, } (L. *scapula'res*), *n* The feathers of the scapular region.
Scap'ularies, } (Plate XI.)
Scar'let (L. *scarlati'nus*), *n* The purest possible red color, lighter and less rosy than carmine, richer and purer than vermilion. (Rose carthame + cadmium orange) (Plate VII fig. 11)
Scar'let-Vermil'ion (L *scarlati'no cinnabari'nus*), *n* Scarlet-vermilion + rose carthame + cadmium orange.) (Plate VII. fig 10)
Schista'ceous (L *schista'ceus*), *n.* Slate-color. (Plate II fig 4)
Schizognath'ous, *a* Having the maxillo-palatine bones separated
Schizorhi'nal, *a* Having the posterior margin of the osseous nares decidedly slit-like or triangular.
Scis'sor-shaped, *a.* A *scissor-shaped tail* is one that is deeply forficate, thus resembling the blades of a pair of shears
Scolo'pacine (L *scolopaci'nus*), *a* Snipe-like Pertaining to or having characteristics of the Snipe family (*Scolopacidæ*)
Scu'tellate (L *scutella'tus*), *a* Provided with scutella, or transverse scales
Scutel'lum (L , pl. *scutel'la*), *n.* One of the regular transverse scales or plates of the tarsus or toes of a bird
Scu'tiform (L *scuti'for'mis*), *a.* Shield-shaped. (Plate XIV fig. 16)
Sea Green (L *thalassi'nus*), *n.* A beautiful deep bluish green color (Italian ultramarine + viridian) (Plate X fig 5)
Seal Brown, *n* A rich, very dark brown color, like the fur of dressed seal-skin. (Lamp-black + vermilion) (Plate III. fig. 1)
Sec'ondary Cov'erts (L *tec'trices seconda'riæ, tec'trices a'læ ma'jores*), *n.* Properly, the posterior row of wing-coverts, which overlie the basal portion of the secondaries. The greater wing-coverts (Plate XI)
Sec'ondaries, } (L *rem'iges seronda'riæ*), *n.* The long feathers of
Sec'ondary Quills, } the forearm, which in the spread wing appear
Sec'ondary Rem'iges, } in a continuous row with the primaries (Plate XI.)
Seg'ment, *n.* A division or specified portion of anything.

Segmenta'tion, *a.* Division into parts or segments.

Semicir'cular, *a* Divided into one half of a circular figure (Plate XIV. fig 20.)

Semilu'nar, *a.* Shaped like a half-moon.

Semipal'mate,
Semipal'mated, } (L. *semipalma'tus*), *a.* Half-webbed, having the membrane between the anterior toes reaching not more than half-way to their ends.

Se'nile (L *seni'lis*), *a* Aged, pertaining to old age.

Se'pia (L.), *n.* A deep dark brown color, with little red in its composition The pigment called sepia is a carbonaceous matter prepared from the natural ink of a species of cuttle-fish. (Plate III. fig. 3.)

Sep'tum (L), *n* A partition

Seri'ceous (L *seri'ceus*), *a* Silky

Ser'rate,
Ser'rated, } (L *serra'tus*), *a.* Toothed like a saw.

Ses'sile (L *ses'silis*), *a.* Resting directly upon an object, without stem, or peduncle

Seta'ceous (L. *seta'ceus*), *a* Bristly, bristled

Se'tæ (L), *n* Bristles, or bristle like feathers.

Se'tiform (L *setifor'mis*), *a.* Bristle-like

Se'vres Blue, *n* A very light blue color. (Antwerp blue + cobalt blue + white.) (Plate IX fig 18.)

Sex'ual, *a* Pertaining to sex.

Shaft (L *rha'chis*), *n* The mid-rib of a feather.

Sib'ilant (L *sibi'lans*), *a* Hissing.

Side of neck (L *parauchen'ium*), *n* The space included between the cervix and the jugulum. (Plate XI.)

Sides, *n* The lateral portions of the inferior surface of a bird's body, extending from near the armpits to, and including, the flanks The sides are subdivisible into (1) sides of breast, (2) sides proper, and (3) flanks (Plate XI.)

Sig'moid, *a* Shaped like the letter S.

Sign (L. *sig'num*), *n* Any character or figure used to denote a word. As, ♂ = *male*; ♀ = *female*; o = *young*; > (in synonymy) = more than, < (in synonymy) = less than, ? = doubt, ! = certainty, etc

Sincip'ital (L *sincipita'lis*), *a* Pertaining to the sinciput, or anterior half of the pileum

Sin'ciput (L), *n* The anterior half of the pileum. (Nearly synonymous with forehead, but denoting a more extensive area, that is, the frontlet, forehead, and anterior part of the crown, together.)

Sin'uate,
Sin'uated, } (L *sinua'tus*), *a* Said of a feather when the edge is gradually cut away. (Plate XIII fig 6.)

Sky Blue (L *azu'reus*, *cœru'leus*; *cœles'tis*, *cœlesti'nus*; *cœlico'lor*), *n* Azure, or light cobalt blue (Same as azure.) (Plate IX fig 15.)

Slate-Black (L. *schista'ceo-ni'ger*), *n.* See plate II. fig. 2.

Slate-co'lor (L. *schista'ceus*), *n.* A dark gray, or blackish gray color, less bluish in tint than plumbeous or lead-color. (Lamp-black + white.) (Plate II fig 4.)

Slate-Gray (L. *schista'ceo-ca'nus*), *n.* (Black + white.) (Plate II fig 5.)

Smalt Blue, *n.* A very deep purplish-blue color, only less intense and rich than hyacinth blue. (Smalt.) (Plate IX fig 8.)

Smoke-Gray (L. *fum'ido-ca'nus*), *n.* (Black + white + raw umber.) (Plate II fig 12.)

Snuff Brown, *n.* A dark brown, essentially the same as a very deep tone of sepia, bistre, or Vandyke brown.

Solferi'no, Solferi'no Pur'ple, *n.* A very beautiful purplish rose-color, intermediate in tint between magenta and rose-red. (Rose aniline + rose tyrien.) (Plate VIII fig 17.)

Spat'ulate (L. *spatula'tus*), *a.* Spoon-shaped, or spatule-shaped, that is, gradually narrowed toward the end, when suddenly widely expanded.

Spe'cies, *n.* The aggregate of individuals related by genetic descent, and differing constantly in certain features whereby they are distinguished from all other beings.

Specif'ic (L. *speci'ficus*), *a.* Pertaining to a species; as, *specific name, specific characters,* etc.

Spec'ulum (L.), *n.* A mirror-like or brightly colored area, usually comprising the secondaries, on the wing of certain ducks.

Spher'ical, *n.* Having the form of a sphere or globe. (Plate XVI fig. 12, representing a section of a sphere.)

Spi'nose, Spi'nous, (L. *spino'sus*), *a.* Having spines, sometimes said of a *mucronate,* or spine-tipped, feather.

Spu'rious (L. *spu'rius*), *a.* False, imperfect; bastard; rudimentary.

Spu'rious Pri'mary, *n.* The first primary, when much reduced in size. (Plate XIII fig 3.)

Spu'rious Wing, *n.* The *alula* or *bastard wing.* (Plate XI.)

Squam'ose, Squam'ous, (L. *squamo'sus*), *a.* Scaly, scale-like, or bearing scales.

Stage, *n.* Used specially for the progressive plumages of birds, as the *immature stage, adult stage, downy stage,* etc. The word *state* is also employed in the same sense.

Steganopo'des (L.), *n.* A group of "Swimming Birds" characterized by having the hind-toe united, on the inner side, to the inner anterior one by a full web. The group includes the Pelicans and allied families. (Same as *Totipalmi.*)

Steganopo'dous, *a.* Having the hallux connected with the anterior toes, as in the Steganopodes.

Stel'late (L. *stella'tus*), *a.* Star-shaped.

Stel'lulate (L. *stellula'tus*), *a.* Resembling little stars.

Ster'ile (L. *ster'ilis*), *a.* Unfruitful; barren.

Stip'ula (L.), *n.* A newly sprouted feather.

Strag'ulum (L.), *n.* The mantle, or the back and upper surface of the wings taken together. (Synonymous with *pallium.*)

Straw-co'lor, ⎰ (L. *strami'neus*), n. A very light impure yellow, like
Straw Yel'low, ⎱ cured straw. (Aureolin + white.) (Plate VI.
fig. 17.)
Streak, n. A narrow longitudinal color-mark; a narrow stripe.
Stri'a (L.; pl. *striæ*), n. A streak.
Stri'ate, ⎱ (L. *stria'tus*), a. Streaked
Stri'ated, ⎰
Stri'dent (L. *stri'dens*), a. Shrill.
Stri'ges (L., plural of *Strix*), n. The name given to the Owl-tribe by those who consider these birds as constituting a distinct order.
Stri'gine, a. Owl-like, pertaining to or having characteristics of the Owl family (*Strigidæ*)
Stripe (L. *pla'ga*), n. A broad, longitudinal color-mark; a broad streak
Struthio'nes (L.), n. The ordinal name of the Ostrich-tribe
Struthio'nine (L. *struthioni'nus*), a. Pertaining to or having characteristics of the Ostrich tribe (*Struthiones*)
Stru'thious, a. Ostrich-like
Sty'liform (L. *styl/for'mis*), ⎱ a. Shaped like a peg or pin.
Sty'loid (L. *styloi'deus*), ⎰
Subarc'uate, ⎱ (L. *subarcua'tus*), a. Slightly arched.
Subarc'uated, ⎰
Sub-ba'sal (L. *subba'salis*), a. Near the base.
Sub-caud'al (L. *subcauda'lis*), a. Beneath the tail
Sub-class (L. *subclas'sis*), n. A group often recognized, having taxonomic rank intermediate between a class and an order
Subfam'ily (L. *subfami'lia*), n. A subdivision of a family including one or more genera
Subge'nus, n. A subdivision of a genus, of indefinite value, and frequently not recognized by name except in the grouping of species
Sub-ma'lar, a. Beneath the *malus*, or malar region, as a sub malar streak. (Plate XII. fig. 20.)
Sub-or'bital, a. Beneath the eye. (Plate XII. fig. 21.)
Sub-or'der (L. *subor'do*), n. A group intermediate in taxonomic rank between an order and a family
Sub-spe'cies, n. A nascent species; a variation, usually geographical, of a species, but not accorded full specific rank on account of the incompleteness of its differentiation; hence, usually a geographical race, or form
Subtyp'ical, a. Not quite typical; somewhat aberrant.
Sub'ulate (L. *subula'tus*), a. Awl-shaped
Suffu'sion, n. A running together of colors.
Sul'cate (L. *sulca'tus*), a. Grooved.
Sul'cus (L.), n. A groove.
Sul'phur Yel'low (L. *sulphu'reus*), n. A very pale pure yellow color, less orange in tint than dilute gamboge or lemon-yellow. (Winsor & Newton's lemon-yellow, or Schoenfeld's gelber ultramarin.) (Plate VI. fig. 14.)

Supercil'iary (L *supercilia'ris*), *a* Above the eye A *superciliary streak*, in its usual sense, denotes a continuous marking of color from the base of the upper mandible over the eye, and extended back above the auriculars to the sides of the occiput. (Plate XII. figs 13, 14, 15, inclusive)

Supercil'ium (L), *n* The eyebrow

Super-fam'ily (L *superfami'lia*), *n* A group containing several families, yet not of ordinal rank (Next in rank below a sub-order.)

Supe'rior, *a* Lying over, topmost, or uppermost

Super-or'der (L. *superor'do*), *n* A group consisting of one or more orders, but not ranking as high as a class (Next in rank below a sub-class.)

Supra-auri'cular (L *supra-auricula'ris*), *a* Situate above the auriculars or ear-coverts (Plate XII fig. 13)

Supralo'ral (L *supralora'lis*), *a* Situate above the lores. (Plate XII. fig 15)

Su'pra-or'bital, *a* Pertaining to the region immediately above the eye. (Plate XII fig 14)

Sym'bol, *n* An arbitrary sign to denote a word. (See *Sign*)

Sym'phesis (L), *n*. A growing together, as *symphesis* of the lower jaw (*symphesis menti*)

Syndac'tyle,
Syndac'tylous, } (L *syndac'tylus*), *a*. Having two toes coalescent for a considerable portion of their length.
Syngne'sious,

Syn'onym,
'Syn'onyme, { (Pl *syn'onyms* or *synon'yma*), *n* In natural history, a specific or generic name which is suppressed on account of having been proposed subsequent to another name for the same object, or for reason of its being otherwise unavailable. Thus, the common Song-sparrow having been first named *fasciata*, in 1788, by Gmelin, the name *melodia*, given by Wilson in 1810, becomes a synonym by reason of its later date. (The opposite of *homonym*, which see)

Synon'ymous, *a* Expressing the same meaning in different terms; or indicating the same genus, species, etc., by a different name.

Synon'ymy, *n*. A collection of synonyms, such as nearly every species is more or less burdened with. The pleasure derived from the study of natural history is seriously marred by the obstacles presented by the synonymy.

Synop'sis, *n* A comprehensive treatment of a given subject, in which only leading points are used

Synop'tical, *a* Pertaining to a synopsis, as a *synoptical table*, in which species or higher groups are distinguished by only the leading characters, arranged antithetically.

Syn'thesis, *n*. Generalization from analyzed facts. (Opposed to *analysis*)

T.

Tail-cov′erts (L *tec′trices-cauda′les*), *n.* The most posterior feathers of the body, or those which immediately cover the basal portion of the tail.

Tar′sal (L *tarsa′lis*), *a.* Pertaining to the tarsus, so-called.

Tar′sus (L), *n.* In descriptive Ornithology, the leg of a bird, or that portion from the foot (that is, the toes) to the heel joint. (Plate XI.)

Taw′ny (L *ful′vus; fulves′cens, aluta′ceus; musteli′nus; fusces′cens*), *a.* The color of tanned leather. (Nearly synonymous with fulvous.) (Neutral orange + raw sienna.) (Plate V. fig 1.)

Taw′ny-ochra′ceous (L *ful′vo-ochra′ceus*), *n.* (Yellow ochre + burnt sienna + burnt umber.) (Plate V fig 4.)

Taw′ny-Ol′ive (L *ful′vo oliva′ceus*), *n.* (Yellow ochre + raw umber.) (Plate III. fig 17.)

Tax′idermist, *n.* A person who prepares and preserves the skins of animals, with the view to imitate their appearance in life.

Taxid′ermy, *n.* The art of preparing and preserving the skins of animals so as to imitate the appearance of life.

Taxonom′ic, *a.* Classificatory, pertaining to taxonomy.

Taxon′omy, *n.* Classification, according to scientific principles.

Tec′trices (L), *n.* Coverts, especially those of the wing.

Tec′trices A′læ,
Tec′trices Ala′res } (L), *n.* Wing-coverts.

Tec′trices Ala′res Inferio′res (L), *n.* The under wing-coverts, or those of the under surface of the wing.

Tec′trices Cau′dæ (L.), *n.* Tail-coverts.

Tec′trices Me′diæ,
Tec′trices Perver′sæ, } (L), *n.* The middle wing-coverts.

Teleolog′ical, *a.* Pertaining to teleology. A *teleological character* is a modification resulting from necessity of adaptation to particular ends. Thus, the naked head and other "vulturine" aspects of the Old World Vultures (belonging to the family *Falconidæ*) and those of the New World (*Cathartidæ*) are teleological, inasmuch as their mode of living necessitates in both certain modifications of external structure fitting them to act the part of scavengers, their actual (morphological) structure being very different.

Teleol′ogy, *n.* The science or doctrine of adaptation.

Tem′poral (L *tempora′lis*), *a.* Pertaining to the temples.

Tenuiros′tral (L. *tenuiros′tris*), *a.* Slender-billed. Pertaining to the obsolete group "Tenuirostres."

Tenuiros'tres (L), *n.* An obsolete and exceedingly artificial group of birds embracing various slender billed forms

Te'rete, *a* Cylindrical and tapering, fusiform

Ter'minal (L *termina'lis*), *a* At the end

Ter'minally, *ad* Toward the end

Terminolog'ical, *a.* Pertaining to terminology

Terminol'ogy, *n.* The science of calling things by their right names, according to fixed or scientific principles, hence, essentially synonymous with *Nomenclature*

Terre-verte' Green, *n* A dull green color, like that produced by the pigment of the same name. (Terre-verte) (Plate X. fig. 3)

Ter'tials,
Ter'tiaries, *n* Properly, the inner quills of the wing, growing from the elbow or humerus, and usually more or less concealed (in the closed wing) by the longer scapulars. Frequently, however, the graduated inner secondaries are incorrectly so called, especially when distinguished, as they very often are, by different color, size, or shape. (Plate XI)

Tes'sellated (L *tessella'tus*), *a* Checkered

Testa'ceous (L *testa'ceus*), *n* or *a* (Same as brick-red) (Plate IV fig 11)

Tetradac'tyle (L. *tetradac'tylus*), *a* Four-toed (Most birds are tetradactyle.)

The'ory, *n.* Scientific speculation, based upon inference from established principles

Thorac'ic (L. *thora'cicus*), *a.* Pertaining to the thorax, or chest

Tho'rax (L), *n.* The chest, or breast

Throat (L *Gu'la*), *n* In descriptive Ornithology, the space between the rami of the lower jaw, including also a small portion of the upper part of the foreneck. (Plate XI)

Thy'roid (L. *thyroi'des*), *a* Shield-shaped

Tib'ia (L), *n* In Osteology, the principal bone of the leg, between the knee and the heel; but in descriptive Ornithology, the so-called "thigh," or shin (Plate XI)

Tib'ial (L *tibia'lis*), *a* Pertaining to the tibia

Tile Red (L *testa'ceus*), *n.* (Same as brick red) (Plate IV fig 11.)

Tinamomorph'æ (L), *n* The "Tinamou-form," equivalent to the *Dromæo'gnathæ* of Huxley.

To'mium (L ; pl *to'mia*), *n.* The cutting-edge of the mandibles, that of the upper being the *maxillary tomium*, that of the lower the *mandibular tomium*

Tor'quate (L. *torqua'tus*), *a* Collared

Totipal'mate (L *totipalma'tus*), *a.* Having the hind-toe united to the anterior toes by a web on one side, as in the Pelicans and other *Steganopodes* (Same as *Steganopodous*)

Totipal'mi (L), *n.* (Same as *Steganopodes*, which see)

Tracheopho'næ (L.), *n* The name of a natural group of passerine birds, characterized by having the syrinx placed at the lower end of the trachea.

Transverse' (L. *transver'sus*), *a* Crosswise, or at right angles with the longitudinal axis of the body or feather.

Transverse'ly, *ad* Crosswise

Tridac'tyle (L. *tridac'tylus*), *a.* Three-toed.

Trino'mial, *a* Composed of three names In Biology, a name composed of three terms, — a generic, a specific, and a subspecific

Triv'ial (L. *trivia'lis*), *a* Sometimes said of a *specific* name

Trochi'li (L.), *n* The ordinal or sub-ordinal name of a natural group of "Picariæ," including only the Humming-birds (*Trochilidæ*) By most authors, however, the group is accorded only family rank, and associated with the families *Cypselidæ* and *Caprimulgidæ*, in a so called order *Macrochires*, or *Cypseli*.

Trun'cate (L *trunca'tus*), *a* Cut squarely off

Turkois' Blue,
Turquoise' Blue, ⎱ (L *turco'sus*), *n* The finest possible light blue color, similar to the stone of the same name (Italian ultramarine + Schoenfeld's "lichtblau" + white) (Plate IX. fig 20)

Tylar'i (L., pl.), *n* The pads on the under surface of the toes

Tym'panum (L), *n* Properly, the ear-drum; but also the naked inflatable air-sacs on the neck of some species of Grouse (*Tetraonidæ*)

Type (L *ty'pus*), *n* Of various signification in Ornithology The *type* of a genus is that species from which the generic characters have been taken, or which is specified as the standard; the *type* of a species is the particular specimen from which the species was originally described The *type*, or typical, form of a group is that which answers best to the diagnosis of that group

Typ'ical, *a.* Agreeing closely with the characters assigned to a group, genus, or species.

U.

Ul'na (L), *n* The posterior bone of the forearm.

Ul'nar, *a.* Pertaining to the ulna.

Ultramarine' Blue (L. *ultramari'nus*; *lazuli'nus*), *n* A very pure lovely blue color, like the pigment called ultramarine (Plate IX fig 9)

Um'ber Brown (L *umbri'nus*), *n* The color of the pigment called raw umber (Plate III fig 14.)

Unarm'ed (L *mu'ticus*), *a.* Said of a toe which has no claw, a tarsus, or a wing, which has no spur, etc.

Un′ciform (L. *uncifor′mis*), ⎫
Un′cinate (L *uncina′tus*), ⎭ *a* Hooked.

Un′der Parts (L *gastræ′um*), *n*. The entire lower surface of a bird, from chin to crissum, inclusive (Same as *Lower Parts.*) (See note facing plate XI.)

Un′der Pri′mary-cov′erts, *n* The primary-coverts of the under surface of the wing (Plate XIII fig 2)

Un′der Tail-cov′erts (L. *tec′trices subcau′dales; calypte′ria inferio′ra*), *n* The feathers immediately beneath the tail. (Practically synonymous with *Crissum*) (Plate XI)

Un′der Wing-cov′erts (L *tec′trices suba′lares*), *n*. The coverts of the under surface of the wing Taken collectively, the term *Lining of the Wing*, or *Wing-lining*, is generally used. (Plate XIII fig 1)

Un′dulate, ⎫ (L *undula′tus*), *a*. Marked with wavy lines (Plate XV
Un′dulated, ⎭ fig. 14)

Unguic′ulate (L *unguicula′tus*), *a*. Clawed

Un′guis (L , pl *un′gues*), *n* A claw.

Unip′arous, *a* Producing but one egg, as the Petrels (*Procellaridæ*) and Auks (*Alcidæ*)

Up′per Parts (L *notæ′um*), *n*. The entire upper surface, from forehead to tail inclusive (See note facing plate XI)

Up′per Tail-cov′erts (L. *tec′trices cau′dæ superio′res , calypte′ria superio′ra*), *n* The feathers overlying the base of the tail, — sometimes produced beyond its tip and simulating the true tail, as in the Peacock (*Pavo cristatus*) and Paradise Trogon (*Pharomacrus moccino*).

Uropy′gial (L *uropygia′lis*), *a* Pertaining to the rump

Uropy′gium (L), *n* The rump. (See plate XI)

U-shaped, *a* Having the form of the letter U. (Plate XV. fig. 4.)

V.

Vandyke′ Brown, *n*. A rich deep brown, very similar to burnt umber but rather less reddish (Plate III. fig. 5)

Vane, *n* The whole of a feather excepting the stem.

Vari′etal, *a* Pertaining to or having the characteristics of a variety

Vari′ety (L *vari′etas*), *n* Properly, an individual or unusual and irregular variation from the normal type of form or coloration, as the various breeds or "strains" of domesticated animals But the term is often, though improperly, applied to subspecies, or geographical races

Vent (L *ven′ter*), *n* The anus

Vent′ral (L *ventra′lis*), *a*. Pertaining to the vent

Vent′ral Re′gion (L *re′gio ventra′lis , re′gio anal′is*), *n* The feathers surrounding or immediately adjacent to the vent. (Plate XI)

Ver′digris Green (L *ærugino′sus*), *n* A very pure and rich green color, appreciably more bluish than viridian (Schoenfeld's dark permanent green.) (Plate X fig 11.)

Ver′diter-Blue, *n* A pale greenish blue, like the pigment of the same name (Plate IX fig 22.)

Vermic′ulate, ⎫ (L. *vermicula′tus*), *a* Marked with irregular fine lines,
Vermic′ulated, ⎭ like the tracks of small worms. (Plate XV fig 13.)

Ver′miform (L *vermifor′mis*), *a* Worm-shaped, as a Woodpecker's tongue

Vermil′ion (L *cinnabari′nus*, *cinnabari′no-ru′ber*), *n* A very fine red color, lighter and less rosy than carmine, and not so pure or rich as scarlet (Plate VII fig 8.)

Ver′nal (L *verna′lis*), *a* Pertaining to Spring

Ver′rucose, ⎫ (L *verruco′sus*), *a*. Warty.
Ver′rucous, ⎭

Ver′satile, *a*. Susceptible of being turned either way, reversible as to position

Ver′tex (L), *n* The crown, or central portion of the pileum. (Plate XI.)

Ver′tical (L *vertica′lis*), *a*. Pertaining to the vertex

Vesti′tus (L), *a* or *n*. Clothed, feathered Clothing, or plumage, as *vesti′tus nuptia′lis*, nuptial or breeding plumage.

Vexil′lum (L), *n* The whole of a feather excepting the stem.

Vibris′sa (L ; pl *vibris′sæ*), *n*. A bristly or bristle-tipped feather, such as those about the gape of a bird.

Vina′ceous (L *vina′ceus*), *n* or *a* A brownish pink, or delicate brownish purple color, like wine-dregs, a soft, delicate wine-colored pink or purple (Schoenfeld's Indian red + white.) (Plate IV. fig 17.)

Vina′ceous-Buff (L *vina′ceo-lu′teus*), *n* (Indian red + yellow ochre + white.) (Plate V. fig 15.)

Vina′ceous-Cin′namon (L *vina′ceo-cinnamo′meus*), *n* (Burnt umber + burnt sienna + white.) (Plate IV fig. 15.)

Vina′ceous-Pink (L *vina′ceo-caryophylla′ceus*), *n* (Madder carmine + light red + white.) (Plate IV. fig 21.)

Vina′ceous-Ru′fous (L *vina′ceo-ru′fus*), *n*. (Schoenfeld's Indian red + light red + white.) (Plate IV fig. 14.)

Vi′olet, ⎫ (L *viola′ceus*, *ianthi′nus*), *n* A purplish blue color, like
Viola′ceous, ⎬ the petals of a violet. (Aniline-violet, or mauve.)
⎭ (Plate VIII. fig 10.)

Vires′cent (L *vires′cens*), *a*. Greenish.

Virid′ian Green, *n* A rich bright green color, somewhat similar to grass-green, but much purer (Plate X fig 8.)

Vit′reous (L *vit′reus*), *a* Glassy, or resembling glass

Vit′ta (L.), *n*. A band of color

V-shaped, *a*. Having the form of the letter V. (Plate XV. fig. 3.)

W.

Wal'nut Brown, *n* A deep warm brown color, like heart-wood of the black walnut. (Sepia + purple madder.) (Plate III fig 7)

War'bler Green, *n* (See *Olive-Green*) (Plate X. fig 18)

Wash'ed (L *afflā'tus, perfū'sus, lavā'tus*), *a* Thinly overlaid with a different color.

Wat'tle (L *pā'lea; verrū'ca*), *n* A pendulous, somewhat fleshy cutaneous flap, usually brightly colored, and often more or less wrinkled, as the "dewlap" of a turkey and the "gills" of the domestic cock

Wa'ved (L *undulā'tus*), *a* Marked with narrow undulating lines of color

Wax Yel'low (L *cerā'ceus*), *n* A deep but dull yellow, resembling the color of fresh bees-wax (Winsor & Newton's "aureolin," or Schoenfeld's "gelber krapplack") (Plate VI fig 7.)

Web (L *pogō'nium*). *n.* Either lateral half of the vane of a feather, exclusive of the shaft.

Wedge-shaped (L. *cuneā'tus*), *a* A *wedge-shaped* tail has the middle pair of feathers longest, the rest successively and decidedly shorter, all more or less attenuate (Plate XIV. fig 13)

Whis'kered (L *mystacā'lis; barbā'tus*), *a* Ornamented by lengthened feathers on the malar region or contiguous portions of the head

Wine Pur'ple (L. *vinā'ceo purpū'reus*), *n* A clear reddish purple of a slightly brownish cast (Madder carmine + violet madder) (Plate VIII fig 15.)

Wood Brown, *n* A light brown color, like some varieties of wood (Raw umber + burnt sienna + white) (Plate III fig 19)

X.

Xiph'oid, *a* Sword-shaped.

Y.

Yel'low O'chre, *n.* A bright yellowish ochraceous or ochre-yellow color (Yellow ochre.) (Plate V fig. 9.)

Z.

Zone (L. *zo′nus*), *n.* A broad band of color, completely encircling the circumference of a body

Zoölog′ical, *a* Pertaining to zoology.

Zoöl′ogy, *n* The natural history of animals in general

Zygodac′tylæ (L.), *n* A group of zygodactyle birds comprising the families *Rhamphastidæ* (Toucans), *Capitonidæ* (Barbets), *Bucconidæ* (Puff-birds), and *Galbulidæ* (Jacamars) In obsolete systems the group was much more extensive, embracing all yoke-footed birds, which are now divided in several distinct groups, *e. g* , the *Pici* (Woodpeckers and Wrynecks), *Anisodactylæ* (Motmots, Todies, Kingfishers, etc.), and *Coccyges* (Cuckoos and Plaintain-eaters)

Zygodac′tyle (L *zygodac′tylus*), *a.* Yoke-toed, or with the toes in pairs, two before and two behind, as in the Woodpeckers, Parrots, etc Pertaining to the Zygodactylæ.

TABLE

FOR CONVERTING MILLIMETRES INTO ENGLISH INCHES AND DECIMALS.

EXPLANATION.

The table herewith given shows the equivalents in English inches, and decimals thereof, of every tenth of a millimetre, from 1.0 to 100.9. From 100 to 1,000 millimetres may be reduced to inches and decimals by multiplying the corresponding figures of this table by ten, that is, by moving the point in the column of inches one place to the right. In a similar way, if multiplying by 100, move the point two places to the right. For example: —

(1) To reduce 72.7 millimetres to English inches: Find in the vertical column to the left the figures 72, then follow the horizontal line to the column headed .7. The number found there (or, what would be sufficient for all practical purposes, 2.86) is the equivalent of 72.7 millimetres in English inches.

(2) To reduce 605 millimetres to English inches, find in the vertical column to the left 60, then follow the line to the column headed .5, where will be found 2.3819; move the point to the right, and you will have as the equivalent of 605 millimetres 23 819 (or, what is essentially the same, 23 82) inches.

(3) To reduce 1930 millimetres, find in the same way the equivalent of 19 3, which is 0.7599; move the point two places to the right, and 75.99 results, which expresses exactly the equivalent of 1930 millimetres in English inches.

TABLE FOR THE REDUCTION OF MILLIMETRES TO ENGLISH INCHES.

Millimetres.	.0	.1	.2	.3	.4	.5	.6	.7	.8	.9
1	0.0397	0.0431	0.0472	0.0511	0.0551	0.0590	0.0629	0.0669	0.0708	0.0748
2	0.0787	0.0826	0.0866	0.0905	0.0944	0.0984	0.1023	0.1062	0.1102	0.1141
3	0.1181	0.1220	0.1259	0.1299	0.1338	0.1378	0.1417	0.1456	0.1496	0.1535
4	0.1575	0.1614	0.1654	0.1693	0.1732	0.1772	0.1811	0.1850	0.1890	0.1929
5	0.1968	0.2008	0.2047	0.2087	0.2126	0.2165	0.2205	0.2244	0.2284	0.2323
6	0.2362	0.2402	0.2441	0.2480	0.2520	0.2559	0.2598	0.2638	0.2677	0.2717
7	0.2756	0.2795	0.2835	0.2874	0.2913	0.2953	0.2992	0.3032	0.3071	0.3110
8	0.3150	0.3189	0.3229	0.3268	0.3307	0.3347	0.3386	0.3425	0.3465	0.3504
9	0.3544	0.3583	0.3622	0.3662	0.3701	0.3740	0.3780	0.3819	0.3859	0.3898
10	0.3937	0.3976	0.4016	0.4055	0.4095	0.4134	0.4173	0.4213	0.4252	0.4291
11	0.4331	0.4370	0.4410	0.4449	0.4488	0.4528	0.4567	0.4606	0.4646	0.4685
12	0.4724	0.4764	0.4803	0.4843	0.4882	0.4921	0.4961	0.5000	0.5040	0.5079
13	0.5118	0.5158	0.5197	0.5236	0.5276	0.5315	0.5354	0.5394	0.5433	0.5472
14	0.5512	0.5551	0.5591	0.5630	0.5669	0.5709	0.5748	0.5788	0.5827	0.5866
15	0.5906	0.5945	0.5984	0.6024	0.6063	0.6102	0.6142	0.6181	0.6221	0.6260
16	0.6299	0.6339	0.6378	0.6417	0.6457	0.6496	0.6536	0.6575	0.6614	0.6654
17	0.6693	0.6732	0.6772	0.6811	0.6850	0.6890	0.6929	0.6969	0.7008	0.7047
18	0.7087	0.7126	0.7165	0.7205	0.7244	0.7284	0.7323	0.7362	0.7402	0.7441
19	0.7480	0.7520	0.7559	0.7599	0.7638	0.7677	0.7717	0.7756	0.7796	0.7835
20	0.7874	0.7913	0.7953	0.7992	0.8032	0.8071	0.8110	0.8150	0.8189	0.8228
21	0.8268	0.8307	0.8347	0.8386	0.8425	0.8465	0.8504	0.8543	0.8583	0.8622
22	0.8662	0.8701	0.8740	0.8780	0.8819	0.8858	0.8898	0.8937	0.8977	0.9016
23	0.9055	0.9095	0.9134	0.9173	0.9213	0.9252	0.9291	0.9331	0.9370	0.9409
24	0.9449	0.9488	0.9528	0.9567	0.9606	0.9646	0.9685	0.9725	0.9764	0.9803
25	0.9843	0.9882	0.9921	0.9961	1.0000	1.0040	1.0079	1.0118	1.0158	1.0197

REDUCTION OF MILLIMETRES TO INCHES. 121

	0	1	2	3	4	5	6	7	8	9
26	1.0236	1.0276	1.0315	1.0355	1.0394	1.0433	1.0473	1.0512	1.0551	1.0591
27	1.0630	1.0669	1.0709	1.0748	1.0788	1.0827	1.0866	1.0906	1.0945	1.0984
28	1.1023	1.1063	1.1102	1.1142	1.1181	1.1221	1.1260	1.1299	1.1339	1.1378
29	1.1418	1.1457	1.1496	1.1536	1.1575	1.1614	1.1654	1.1693	1.1732	1.1772
30	1.1811	1.1851	1.1890	1.1929	1.1969	1.2008	1.2047	1.2087	1.2126	1.2166
31	1.2205	1.2244	1.2284	1.2323	1.2362	1.2402	1.2441	1.2481	1.2520	1.2559
32	1.2599	1.2638	1.2677	1.2717	1.2756	1.2795	1.2835	1.2874	1.2914	1.2953
33	1.2992	1.3032	1.3071	1.3110	1.3150	1.3189	1.3228	1.3268	1.3307	1.3347
34	1.3386	1.3425	1.3465	1.3504	1.3544	1.3583	1.3622	1.3662	1.3701	1.3740
35	1.3780	1.3819	1.3859	1.3898	1.3935	1.3977	1.4016	1.4055	1.4095	1.4134
36	1.4173	1.4213	1.4252	1.4292	1.4331	1.4370	1.4410	1.4449	1.4488	1.4528
37	1.4567	1.4607	1.4646	1.4685	1.4725	1.4764	1.4803	1.4843	1.4882	1.4922
38	1.4961	1.5000	1.5040	1.5079	1.5118	1.5158	1.5197	1.5236	1.5276	1.5315
39	1.5355	1.5394	1.5433	1.5473	1.5512	1.5551	1.5591	1.5630	1.5670	1.5709
40	1.5748	1.5788	1.5827	1.5866	1.5906	1.5945	1.5985	1.6024	1.6063	1.6103
41	1.6142	1.6181	1.6221	1.6260	1.6299	1.6339	1.6378	1.6418	1.6458	1.6496
42	1.6536	1.6575	1.6614	1.6654	1.6693	1.6733	1.6772	1.6811	1.6851	1.6890
43	1.6929	1.6969	1.7008	1.7048	1.7087	1.7126	1.7166	1.7205	1.7244	1.7284
44	1.7323	1.7362	1.7402	1.7441	1.7481	1.7520	1.7559	1.7599	1.7638	1.7677
45	1.7717	1.7756	1.7796	1.7835	1.7874	1.7914	1.7953	1.7992	1.8032	1.8071
46	1.8111	1.8150	1.8189	1.8228	1.8268	1.8307	1.8347	1.8386	1.8425	1.8465
47	1.8504	1.8543	1.8583	1.8622	1.8662	1.8701	1.8740	1.8780	1.8819	1.8858
48	1.8898	1.8937	1.8977	1.9016	1.9055	1.9095	1.9134	1.9173	1.9213	1.9252
49	1.9291	1.9331	1.9370	1.9409	1.9449	1.9488	1.9528	1.9567	1.9606	1.9646
50	1.9685	1.9725	1.9764	1.9803	1.9843	1.9882	1.9921	1.9961	2.0000	2.0040
51	2.0079	2.0118	2.0158	2.0197	2.0236	2.0276	2.0315	2.0355	2.0394	2.0433
52	2.0473	2.0512	2.0551	2.0591	2.0630	2.0669	2.0709	2.0748	2.0788	2.0827
53	2.0866	2.0906	2.0945	2.0984	2.1023	2.1063	2.1102	2.1142	2.1181	2.1221
54	2.1260	2.1299	2.1339	2.1378	2.1418	2.1457	2.1496	2.1536	2.1575	2.1614
55	2.1654	2.1693	2.1732	2.1772	2.1811	2.1851	2.1890	2.1929	2.1969	2.2008

TABLE FOR THE REDUCTION OF MILLIMETRES TO ENGLISH INCHES. — *Continued*

MILLIMETRES.	.0	.1	.2	.3	.4	.5	.6	.7	.8	.9
56	2.2047	2.2087	2.2126	2.2166	2.2205	2.2244	2.2284	2.2323	2.2362	2.2402
57	2.2441	2.2481	2.2520	2.2559	2.2599	2.2638	2.2677	2.2717	2.2756	2.2795
58	2.2835	2.2874	2.2914	2.2953	2.2992	2.3032	2.3071	2.3110	2.3150	2.3189
59	2.3229	2.3268	2.3307	2.3347	2.3386	2.3425	2.3465	2.3504	2.3544	2.3583
60	2.3622	2.3662	2.3701	2.3740	2.3780	2.3819	2.3858	2.3898	2.3937	2.3977
61	2.4016	2.4055	2.4095	2.4134	2.4173	2.4213	2.4252	2.4292	2.4331	2.4370
62	2.4410	2.4449	2.4488	2.4528	2.4567	2.4607	2.4646	2.4685	2.4725	2.4764
63	2.4803	2.4843	2.4882	2.4922	2.4961	2.5000	2.5040	2.5079	2.5118	2.5158
64	2.5197	2.5236	2.5276	2.5315	2.5355	2.5394	2.5433	2.5473	2.5512	2.5551
65	2.5591	2.5630	2.5670	2.5709	2.5748	2.5788	2.5827	2.5866	2.5906	2.5945
66	2.5985	2.6024	2.6063	2.6103	2.6142	2.6181	2.6221	2.6260	2.6300	2.6339
67	2.6378	2.6418	2.6458	2.6496	2.6536	2.6575	2.6614	2.6654	2.6693	2.6733
68	2.6772	2.6811	2.6851	2.6890	2.6929	2.6969	2.7008	2.7048	2.7087	2.7126
69	2.7166	2.7205	2.7244	2.7284	2.7323	2.7362	2.7402	2.7441	2.7481	2.7520
70	2.7559	2.7599	2.7636	2.7677	2.7717	2.7756	2.7796	2.7835	2.7874	2.7914
71	2.7953	2.7992	2.8032	2.8071	2.8111	2.8150	2.8189	2.8228	2.8268	2.8307
72	2.8347	2.8386	2.8425	2.8465	2.8504	2.8543	2.8583	2.8622	2.8662	2.8701
73	2.8740	2.8780	2.8819	2.8858	2.8898	2.8937	2.8977	2.9016	2.9055	2.9095
74	2.9134	2.9173	2.9213	2.9252	2.9291	2.9331	2.9370	2.9409	2.9449	2.9488
75	2.9528	2.9567	2.9606	2.9646	2.9685	2.9725	2.9764	2.9803	2.9843	2.9882
76	2.9921	2.9961	3.0000	3.0040	3.0079	3.0118	3.0158	3.0197	3.0236	3.0276
77	3.0315	3.0355	3.0394	3.0433	3.0473	3.0512	3.0551	3.0591	3.0630	3.0669
78	3.0709	3.0748	3.0788	3.0827	3.0866	3.0906	3.0945	3.0984	3.1023	3.1063
79	3.1103	3.1142	3.1181	3.1221	3.1260	3.1299	3.1339	3.1378	3.1418	3.1457
80	3.1496	3.1536	3.1575	3.1614	3.1654	3.1693	3.1732	3.1772	3.1811	3.1851

REDUCTION OF MILLIMETRES TO INCHES.

	0	1	2	3	4	5	6	7	8	9
81	3.1890	3.1929	3.1969	3.2008	3.2047	3.2087	3.2126	3.2166	3.2205	3.2244
82	3.2284	3.2323	3.2362	3.2402	3.2441	3.2481	3.2520	3.2559	3.2599	3.2638
83	3.2677	3.2717	3.2756	3.2795	3.2835	3.2874	3.2914	3.2953	3.2992	3.3032
84	3.3071	3.3110	3.3150	3.3189	3.3229	3.3268	3.3307	3.3347	3.3386	3.3425
85	3.3465	3.3504	3.3544	3.3583	3.3622	3.3662	3.3701	3.3740	3.3780	3.3819
86	3.3859	3.3898	3.3937	3.3977	3.4016	3.4055	3.4095	3.4134	3.4173	3.4213
87	3.4252	3.4292	3.4331	3.4370	3.4410	3.4449	3.4488	3.4528	3.4567	3.4607
88	3.4646	3.4685	3.4725	3.4764	3.4803	3.4843	3.4882	3.4922	3.4961	3.5000
89	3.5040	3.5079	3.5118	3.5158	3.5197	3.5236	3.5276	3.5315	3.5355	3.5394
90	3.5433	3.5473	3.5512	3.5551	3.5591	3.5630	3.5670	3.5709	3.5748	3.5788
91	3.5827	3.5866	3.5906	3.5945	3.5985	3.6024	3.6063	3.6103	3.6142	3.6181
92	3.6221	3.6260	3.6300	3.6339	3.6378	3.6418	3.6458	3.6496	3.6536	3.6575
93	3.6614	3.6654	3.6693	3.6733	3.6772	3.6811	3.6850	3.6890	3.6929	3.6969
94	3.7008	3.7048	3.7087	3.7126	3.7166	3.7205	3.7244	3.7284	3.7323	3.7362
95	3.7402	3.7441	3.7481	3.7520	3.7559	3.7599	3.7638	3.7677	3.7717	3.7756
96	3.7796	3.7835	3.7874	3.7914	3.7953	3.7992	3.8032	3.8071	3.8111	3.8150
97	3.8189	3.8228	3.8268	3.8307	3.8347	3.8386	3.8425	3.8465	3.8504	3.8543
98	3.8583	3.8622	3.8662	3.8701	3.8740	3.8780	3.8819	3.8858	3.8898	3.8937
99	3.8977	3.9016	3.9055	3.9095	3.9134	3.9173	3.9213	3.9252	3.9291	3.9331
100	3.9370	3.9409	3.9449	3.9488	3.9528	3.9567	3.9606	3.9646	3.9685	3.9725

TABLE

FOR CONVERTING ENGLISH INCHES AND DECIMALS INTO MILLIMETRES.

———◆———

EXPLANATION.

The accompanying table shows the equivalent of every hundredth of an English inch, from 0 01 to 10 09 inches, in millimetres and decimals thereof. From 10 to 100 inches may be reduced to millimetres and decimals by multiplying the corresponding figures of this table by 10; that is, by moving the point in the column of inches one place to the right. In a similar way, if multiplying by 100, move the point two places to the right. For example. —

(1) To reduce 4.36 inches to millimetres: Find in the first or left-hand column the figures 4.3; then follow the horizontal line toward the right to the column headed .06 The number found there, 110.74 (or, what would be sufficiently near for all practical purposes, 111), is the equivalent of 4 36 inches in millimetres.

(2) To reduce 15 77 inches to millimetres. Find in the first or left-hand column the figures 1 5; then follow the horizontal line toward the right to the column headed .07, where will be found the figures 39 88; now move the point one place to the right, and you have 398 8, which is the exact equivalent of 15 70 inches; now find the equivalent of 0.07 inches, which is 1.78 millimetres, and add to the

REDUCTION OF INCHES TO MILLIMETRES.

above amount, the result being 400 58 millimetres (or, what is sufficiently near, 401 millimetres), which = 15.77 inches.

(3) To reduce 120.44 inches to millimetres, find in the same manner the equivalent of 1 2, which is 30 48, move the point two places to the right, and 3048. results; add the equivalent of 0 44, which is 11.18, and the result is 3059 millimetres = 120.44 inches.

TABLE FOR THE REDUCTION OF ENGLISH INCHES TO MILLIMETRES.

Inches.	.00	.01	.02	.03	.04	.05	.06	.07	.08	.09
0.0	0.00	0.25	0.51	0.76	1.02	1.27	1.52	1.78	2.03	2.29
0.1	2.54	2.79	3.05	3.30	3.56	3.81	4.06	4.32	4.57	4.83
0.2	5.08	5.33	5.59	5.84	6.10	6.35	6.60	6.86	7.11	7.37
0.3	7.62	7.87	8.13	8.38	8.64	8.89	9.14	9.40	9.65	9.91
0.4	10.16	10.41	10.67	10.92	11.18	11.43	11.68	11.94	12.19	12.45
0.5	12.70	12.95	13.21	13.46	13.72	13.97	14.22	14.48	14.73	14.99
0.6	15.24	15.49	15.75	16.00	16.26	16.51	16.76	17.02	17.27	17.53
0.7	17.78	18.03	18.29	18.54	18.80	19.05	19.30	19.56	19.81	20.07
0.8	20.32	20.57	20.83	21.08	21.34	21.59	21.84	22.10	22.35	22.61
0.9	22.86	23.11	23.37	23.62	23.88	24.13	24.38	24.64	24.89	25.15
1.0	25.40	25.65	25.91	26.16	26.42	26.67	26.92	27.18	27.43	27.68
1.1	27.94	28.19	28.45	28.70	28.96	29.21	29.46	29.72	29.97	30.22
1.2	30.48	30.73	30.99	31.24	31.50	31.75	32.00	32.26	32.51	32.76
1.3	33.02	33.27	33.53	33.78	34.04	34.29	34.54	34.80	35.05	35.30
1.4	35.56	35.81	36.07	36.32	36.58	36.83	37.08	37.34	37.59	37.84
1.5	38.10	38.35	38.61	38.86	39.12	39.37	39.62	39.88	40.13	40.38
1.6	40.64	40.89	41.15	41.40	41.66	41.91	42.16	42.42	42.67	42.92
1.7	43.18	43.43	43.69	43.94	44.20	44.45	44.70	44.96	45.21	45.46
1.8	45.72	45.97	46.23	46.48	46.74	46.99	47.24	47.50	47.75	48.00
1.9	48.26	48.51	48.77	49.02	49.28	49.53	49.78	50.04	50.29	50.54
2.0	50.80	51.05	51.31	51.56	51.82	52.07	52.32	52.58	52.83	53.08
2.1	53.34	53.59	53.85	54.10	54.36	54.61	54.86	55.12	55.37	55.62
2.2	55.88	56.13	56.39	56.64	56.90	57.15	57.40	57.66	57.91	58.16
2.3	58.42	58.67	58.93	59.18	59.44	59.69	59.94	60.20	60.45	60.70
2.4	60.96	61.21	61.47	61.72	61.97	62.23	62.48	62.74	62.99	63.24
2.5	63.50	63.75	64.01	64.26	64.52	64.77	65.02	65.28	65.53	65.78

REDUCTION OF INCHES TO MILLIMETRES. 127

	.0	.1	.2	.3	.4	.5	.6	.7	.8	.9
2.6	66.04	66.29	66.55	66.80	67.06	67.31	67.56	67.82	68.07	68.32
2.7	68.58	68.83	69.09	69.34	69.60	69.85	70.10	70.36	70.61	70.86
2.8	71.12	71.37	71.63	71.88	72.14	72.39	72.64	72.90	73.15	73.40
2.9	73.66	73.91	74.17	74.42	74.68	74.93	75.18	75.44	75.69	75.94
3.0	76.20	76.45	76.71	76.96	77.21	77.47	77.72	77.98	78.23	78.48
3.1	78.74	78.99	79.25	79.50	79.75	80.01	80.26	80.52	80.77	81.02
3.2	81.28	81.53	81.79	82.04	82.29	82.55	82.80	83.06	83.31	83.56
3.3	83.82	84.07	84.33	84.58	84.83	85.09	85.34	85.60	85.85	86.10
3.4	86.36	86.61	86.87	87.12	87.37	87.63	87.88	88.14	88.39	88.64
3.5	88.90	89.15	89.41	89.66	89.91	90.17	90.42	90.68	90.93	91.18
3.6	91.44	91.69	91.95	92.20	92.45	92.71	92.96	93.22	93.47	93.72
3.7	93.98	94.23	94.49	94.74	94.99	95.25	95.50	95.75	96.01	96.26
3.8	96.52	96.77	97.02	97.28	97.53	97.79	98.04	98.29	98.55	98.80
3.9	99.05	99.31	99.56	99.82	100.07	100.33	100.58	100.83	101.09	101.34
4.0	101.60	101.85	102.10	102.36	102.61	102.87	103.12	103.37	103.63	103.88
4.1	104.14	104.39	104.64	104.90	105.15	105.41	105.66	105.91	106.17	106.42
4.2	104.68	104.93	107.18	107.44	107.69	107.95	108.20	108.45	108.71	108.96
4.3	109.22	109.47	109.72	109.98	110.23	110.49	110.74	110.99	111.25	111.50
4.4	111.76	112.01	112.26	112.52	112.77	113.02	113.28	113.53	113.79	114.04
4.5	114.30	114.55	114.80	115.06	115.31	115.56	115.82	116.07	116.33	116.58
4.6	116.84	117.09	117.34	117.60	117.85	118.10	118.36	118.61	118.87	119.12
4.7	119.38	119.63	119.88	120.14	120.39	120.64	120.90	121.15	121.41	121.66
4.8	121.92	122.17	122.42	122.68	122.93	123.18	123.44	123.69	123.95	124.20
4.9	124.46	124.71	124.96	125.22	125.47	125.72	125.98	126.23	126.49	126.74
5.0	127.00	127.25	127.50	127.76	128.01	128.26	128.52	128.77	129.02	129.28
5.1	129.54	129.79	130.04	130.30	130.55	130.80	131.06	131.31	131.56	131.82
5.2	132.08	132.33	132.58	132.84	133.09	133.34	133.60	133.85	134.10	134.36
5.3	134.62	134.87	135.12	135.38	135.63	135.88	136.14	136.39	136.64	136.90
5.4	137.16	137.41	137.66	137.92	138.17	138.42	138.68	138.93	139.18	139.44
5.5	138.70	139.95	140.20	140.46	140.71	140.96	141.22	141.47	141.72	141.98

REDUCTION OF INCHES TO MILLIMETRES.

Table for the Reduction of English Inches to Millimetres. — *Continued.*

Inches.	.00	.01	.02	.03	.04	.05	.06	.07	.08	.09
5.6	142.24	142.49	142.75	143.00	143.26	143.51	143.76	144.01	144.26	144.52
5.7	144.78	145.03	145.28	145.54	145.79	146.05	146.30	146.56	146.81	147.07
5.8	147.32	147.57	147.83	148.08	148.34	148.59	148.84	149.10	149.35	149.61
5.9	149.86	150.11	150.37	150.62	150.88	151.13	151.38	151.64	151.89	152.15
6.0	152.40	152.65	152.91	153.16	153.42	153.67	153.92	154.18	154.43	154.68
6.1	154.94	155.19	155.45	155.70	155.96	156.21	156.46	156.72	156.97	157.23
6.2	157.48	157.73	157.99	158.24	158.50	158.75	159.00	159.26	159.51	159.76
6.3	160.02	160.27	160.53	160.78	161.04	161.29	161.54	161.80	162.05	162.30
6.4	162.56	162.81	163.07	163.32	163.58	163.83	164.08	164.34	164.59	164.84
6.5	165.10	165.35	165.61	165.86	166.12	166.37	166.62	166.88	167.13	167.38
6.6	167.64	167.89	168.15	168.40	168.66	168.91	169.16	169.42	169.67	169.92
6.7	170.18	170.43	170.69	170.94	171.20	171.45	171.70	171.96	172.21	172.46
6.8	172.72	172.97	173.23	173.48	173.74	173.99	174.24	174.50	174.75	175.00
6.9	175.26	175.51	175.77	176.02	176.28	176.53	176.78	177.04	177.29	177.54
7.0	177.80	178.05	178.31	178.56	178.82	179.07	179.32	179.58	179.83	180.08
7.1	180.34	180.59	180.85	181.10	181.36	181.61	181.86	182.12	182.37	182.62
7.2	182.88	183.13	183.39	183.64	183.90	184.15	184.40	184.66	184.91	185.16
7.3	185.42	185.67	185.93	186.18	186.44	186.69	186.94	187.20	187.45	187.70
7.4	187.96	188.21	188.47	188.72	188.98	189.23	189.48	189.74	189.99	190.24
7.5	190.50	190.75	191.01	191.26	191.52	191.77	192.02	192.28	192.53	192.78
7.6	193.04	193.29	193.55	193.80	194.06	194.31	194.56	194.82	195.07	195.32
7.7	195.58	195.83	196.09	196.34	196.60	196.85	197.10	197.36	197.61	197.86
7.8	198.12	198.37	198.63	198.88	199.14	199.39	199.64	199.90	200.15	200.40
7.9	200.66	200.91	201.17	201.42	201.68	201.93	202.18	202.44	202.69	202.94
8.0	203.20	203.45	203.71	203.96	204.21	204.47	204.74	204.98	205.23	205.48

REDUCTION OF INCHES TO MILLIMETRES.

in.	.00	.01	.02	.03	.04	.05	.06	.07	.08	.09
8.1	205.74	205.99	206.25	206.50	206.75	207.01	207.26	207.52	207.77	208.02
8.2	208.28	208.53	208.79	209.04	209.29	209.55	209.80	210.06	210.31	210.56
8.3	210.82	211.07	211.33	211.58	211.83	212.09	212.31	212.60	212.85	213.10
8.4	213.36	213.61	213.87	214.12	214.37	214.63	214.88	215.14	215.39	215.64
8.5	215.90	216.15	216.41	216.66	216.91	217.17	217.42	217.68	217.93	218.18
8.6	218.44	218.69	218.95	219.20	219.45	219.71	219.96	220.22	220.47	220.72
8.7	220.98	221.23	221.49	221.74	221.99	222.25	222.50	222.75	223.01	223.26
8.8	223.52	223.77	224.02	224.28	224.53	224.79	225.04	225.29	225.55	225.80
8.9	226.05	226.31	226.56	226.82	227.07	227.33	227.58	227.83	228.09	228.34
9.0	228.60	228.85	229.10	229.36	229.61	229.87	230.12	230.37	230.63	230.88
9.1	231.14	231.39	231.64	231.90	232.15	232.41	232.66	232.91	233.17	233.42
9.2	233.68	233.93	234.18	234.44	234.69	234.95	235.20	235.45	235.71	235.96
9.3	236.22	236.47	236.72	236.98	237.23	237.49	237.74	237.99	238.25	238.50
9.4	238.76	239.01	239.26	239.52	239.77	240.02	240.28	240.53	240.79	241.04
9.5	241.30	241.55	241.80	242.06	242.31	242.56	242.82	243.07	243.33	243.58
9.6	243.84	244.09	244.34	244.60	244.85	245.10	245.36	245.61	245.87	246.12
9.7	246.38	246.63	246.88	247.14	247.39	247.64	247.90	248.15	248.41	248.66
9.8	248.92	249.17	249.42	249.68	249.93	250.18	250.44	250.69	250.95	251.20
9.9	251.46	251.71	251.96	252.22	252.47	252.72	252.98	253.23	253.49	253.74
10.0	254.00	254.25	254.50	254.76	255.01	255.26	255.52	255.77	256.02	256.28

PLATE I.

I — PRIMARY COMBINATIONS

The primaries here used are —
    Yellow — light cadmium.
    Red — scarlet vermilion + madder carmine.
    Blue — ultramarine

The secondaries, however, are not all of them produced by the mixture of the pigments used as primaries, the orange being *orange cadmium*, and the purple, *aniline violet*. This is necessary on account of the impurity of the red

---

II. — SECONDARY COMBINATIONS

The pure secondaries in this figure are the same as those in the one above. The tertiaries, however, are here in each case *actual mixtures* of the pigments

I.— Primary Combinations.

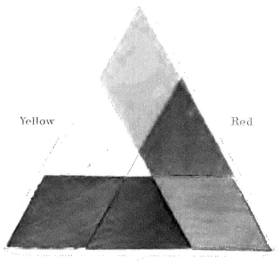

Yellow

Red

Blue

II.— Secondary Combinations.

Purple

Orange

Green

PLATE II

Composed of

1. Black — *Lamp-black*
2. Plumbeous . Lamp-black + white + smalt blue
3. Olive-Gray . . Lamp-black + white + light cadmium
4. Slate-color . . Lamp-black + white.
5. Cinereous . . Lamp-black + white + smalt blue.
6. Mouse Gray . Lamp-black + white + sepia.
7. Slate-Gray . . Lamp-black + white.
8. Lilac Gray — Lamp-black + white + cobalt blue + madder carmine
9. Smoke Gray . Lamp-black + white + raw umber.
10. French Gray . . Lamp black + white + intense blue.
11. Lavender Gray . Lamp-black + white + smalt blue.
12. Drab-Gray . Lamp-black + white + burnt umber
13. Pearl Gray . . Lamp-black + white + cobalt blue.

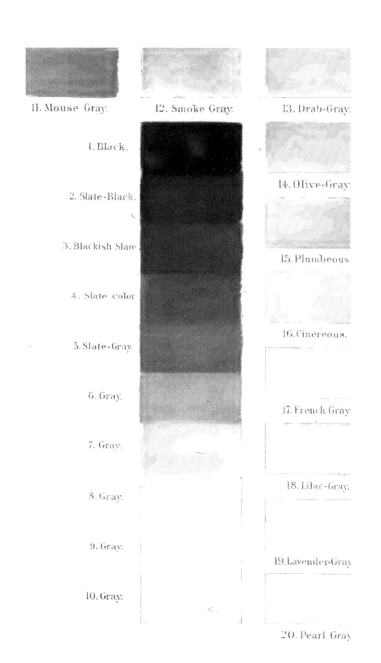

PLATE III

Composed of

1. Seal Brown   Lamp-black + vermilion.
2. Clove Brown .   Lamp-black + cadmium orange
3. Sepia . .   (Sepia)
4. Chocolate Brown   Purple madder + sepia.
5. Vandyke Brown   (Vandyke brown)
6. Bistre . . .   (Bistre)
7. Walnut Brown   Burnt umber + purple madder.
8. Burnt Umber .   (Burnt umber)
9. Olive .   Sepia + yellow ochre + Antwerp blue.
10. Mummy Brown   Raw umber + burnt sienna.
11. Prout's Brown   Raw umber + burnt umber + sepia
12. Hair Brown   Raw umber + sepia + black + white
13. Mars Brown   Burnt umber + yellow ochre + burnt sienna
14. Raw Umber   (Raw umber).
15. Broccoli Brown   Sepia + raw umber + white.
16. Russet .   Burnt umber + burnt sienna + yellow ochre
17. Tawny Olive   Raw umber + yellow ochre.
18. Drab . .   Sepia + white.
19. Wood Brown .   Burnt umber + raw umber + yellow ochre + white.
20. Cinnamon .   Yellow ochre + burnt umber + burnt sienna + white.
21. Ecru-Drab .   Burnt umber + sepia + white
22. Fawn-color .   Burnt umber + white
23. Isabella-color   Raw umber + yellow ochre + white

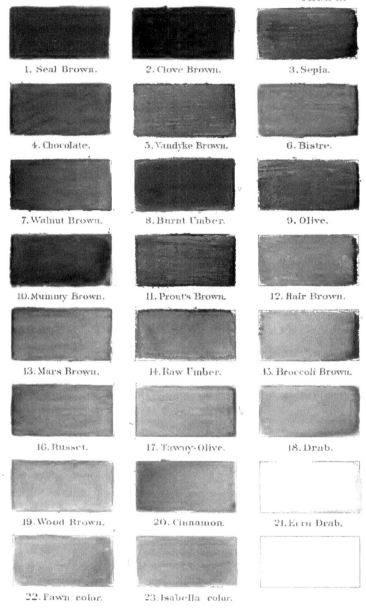

PLATE IV.

Composed of

1. Claret Brown . . Purple madder.
2. Maroon . . . Purple madder + madder carmine + scarlet vermilion.
3. Madder Brown . Purple madder + burnt sienna
4. Liver Brown . . Schoenfeld's Indian red or Persian red
5. Bay . . . Schoenfeld's Indian red + burnt sienna
6. Burnt Sienna . (Burnt sienna)
7. Rufous . . . (Light red).
8. Dragon's-blood Red Light red + vermilion + madder carmine
9. Chestnut . . . . Burnt umber + vermilion
10. Ferruginous . (Burnt sienna, light tint)
11. Brick Red . . Winsor & Newton's Indian red
12. Hazel . . . . . Burnt sienna + vermilion + raw sienna
13. Orange-Rufous . Neutral orange, or light red + orange cadmium
14. Vinaceous Rufous . Persian red + light red + white
15. Vinaceous Cinnamon Burnt umber + burnt sienna + white.
16. Cinnamon-Rufous . Burnt sienna + burnt umber + light red + white
17. Vinaceous . . . Schoenfeld's Indian red + white
18. Pinkish Vinaceous . Winsor & Newton's Indian red + white
19. Salmon-Buff . . Light red + cadmium orange + white.
20. Buff-Pink . Light red + white.
21. Vinaceous Pink . Madder carmine + light red + white.

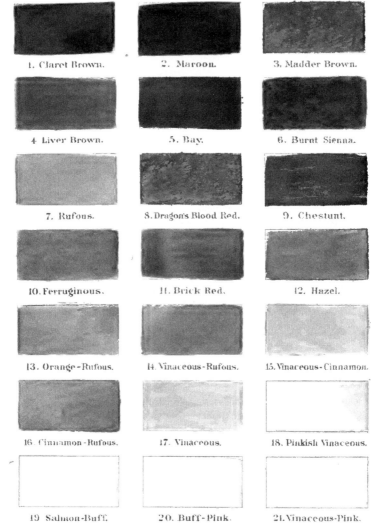

PLATE V.

Composed of

1. Tawny . . . . Raw sienna + burnt sienna
2. Raw Sienna (Raw sienna).
3. Orange-Ochraceous Cadmium orange + yellow ochre + burnt sienna.
4. Tawny Ochraceous Burnt sienna + burnt umber + yellow ochre.
5. Ochraceous Rufous Yellow ochre + light red + burnt sienna
6. Gallstone Yellow Raw sienna + light cadmium.
7. Ochraceous . . Yellow ochre + burnt umber + burnt sienna
8. Clay-color . . . Yellow ochre + raw umber + burnt sienna + white
9. Ochre Yellow . . (Yellow ochre)
10. Ochraceous Buff . Yellow ochre + burnt sienna + white
11. Cream Buff . . Yellow ochre + white
12. Olive-Buff Yellow ochre + white + cobalt blue
13. Buff . . . . Raw sienna + white.
14. Pinkish Buff . Yellow ochre + light red + white
15. Vinaceous Buff Yellow ochre + white + Schoenfeld's Indian red.

PLATE V

1. Tawny.   2. Raw Sienna.   3. Orange-Ochraceous.

4. Tawny Ochraceous.   5. Ochraceous-Rufous.   6. Gallstone Yellow.

7. Ochraceous.   8. Clay color.   9. Ochre Yellow.

10. Ochraceous-Buff.   11. Cream-Buff.   12. Olive-Buff.

13. Buff.   14. Pinkish Buff.   15. Vinaceous-Buff.

PLATE VI.

Composed of

Orpiment Orange . .	Cadmium orange + burnt sienna
Cadmium Orange .	(Cadmium orange).
Orange . .	Orange cadmium + pale cadmium.
Saffron Yellow	Pale cadmium + orange cadmium + raw sienna.
Indian Yellow . . .	Pale cadmium + orange cadmium + yellow ochre
Cadmium Yellow	Pale cadmium + orange cadmium.
Wax Yellow .	Pale cadmium + raw umber.
Chrome Yellow	Pale cadmium + orange cadmium + white
Deep Chrome . .	Pale cadmium + orange cadmium + white
Gamboge Yellow . .	Pale cadmium + yellow ochre
Lemon Yellow . .	Pale cadmium.
Canary Yellow . .	Pale cadmium + white.
Primrose Yellow	Pale cadmium + white.
Sulphur Yellow . .	Pale cadmium + white + Paris green
Citron Yellow	Pale cadmium + Antwerp blue
Olive Yellow	Pale cadmium + black + white
Straw Yellow	Pale cadmium + raw umber + white
Naples Yellow .	Pale cadmium + yellow ochre + white.
Buff Yellow .	Pale cadmium + orange cadmium + white
Cream-color . . .	Orange cadmium + pale cadmium + white.
Maize Yellow	Orange cadmium + pale cadmium + white
Orange-Buff . . .	Orange cadmium + white

1. Orpiment Orange. 2. Cadmium Orange. 3. Orange.

4. Saffron Yellow. 5. Indian Yellow. 6. Cadmium Yellow.

7. Wax Yellow. 8. Chrome Yellow 9. Deep Chrome.

10. Gamboge Yellow 11. Lemon Yellow. 12. Canary Yellow.

13. Primrose Yellow. 14. Sulphur Yellow. 15. Citron Yellow.

16. Olive-Yellow. 17. Straw Yellow. 18. Naples Yellow.

19. Buff-Yellow. 20. Cream color. 21. Maize Yellow.

22. Orange-Buff.

PLATE VII.

	Composed of
Burnt Carmine	Madder carmine + scarlet vermilion + black
Lake Red	Madder carmine or deep madder lake
Crimson	Madder carmine or deep madder lake.
Coral Red	Madder carmine + cadmium orange + vermilion + white.
Rose Red	Bourgeois's "rose-Tyrien" + Schoenfeld's "safflorroth."
Carmine.	
Geranium Red	Schoenfeld's "saffloroth."
Vermilion	Winsor & Newton's vermilion.
Poppy Red	Bourgeois's "laque ponceau."
Scarlet Vermilion	Winsor & Newton's scarlet vermilion.
Scarlet	Saffloroth + cadmium orange.
Orange-Vermilion	Scarlet vermilion + cadmium orange.
Orange Chrome	Vermilion + cadmium orange.
Flame-scarlet	Saffloroth + cadmium orange
Chinese Orange	Scarlet vermilion + cadmium orange + burnt sienna
Saturn Red	Scarlet vermilion + cadmium orange.
Salmon-color	Scarlet vermilion + cadmium orange + white.
Flesh-color	Scarlet vermilion + white.
Geranium Pink	Saffloroth + white.
Rose Pink	Saffloroth + white
Peach-blossom Pink	Pink madder.

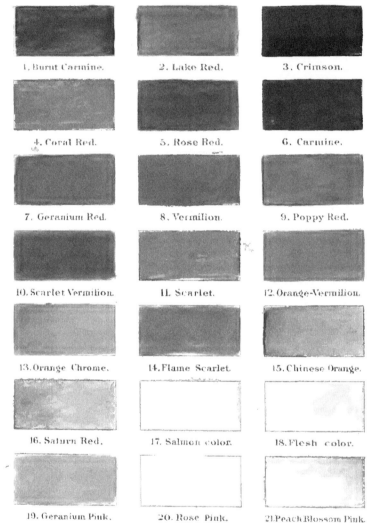

PLATE VIII.

	Composed of
Prune Purple . . .	Violet madder lake + purple madder
Dahlia Purple	Madder carmine + black + violet ultramarine
Auricula Purple	Violet madder lake (Schoenfeld's)
Plum Purple . . .	Violet madder lake + Antwerp blue.
Pansy Purple . .	Madder carmine + violet ultramarine + black
Indian Purple	Violet madder lake + sepia
Royal Purple .	Aniline violet + violet ultramarine
Aster Purple .	Violet madder lake + violet ultramarine + madder carmine.
Maroon Purple	Madder carmine + purple madder
Violet	Aniline violet.
Phlox Purple .	Violet madder lake + violet ultramarine + madder carmine + white
Pomegranate Purple.	Madder carmine + violet madder lake.
Mauve . .	Aniline violet + white
Magenta	Aniline violet + rose aniline
Wine Purple	Madder carmine + violet madder lake
Lavender .	Violet ultramarine + white
Solferino	Rose aniline
Heliotrope	Violet madder lake + sepia + violet ultramarine + white
Lilac	Violet ultramarine + madder carmine + white
Rose Purple	Violet ultramarine + madder carmine + white

PLATE IX.

	Composed of
Indigo Blue . .	Antwerp blue + black
Marine Blue . . .	French blue + black.
———— .	French blue + violet ultramarine.
Berlin Blue .	Antwerp blue + French blue + black
Hyacinth Blue .	Schoenfeld's violet ultramarine.
French Blue .	(French blue).
Paris Blue . .	Antwerp blue + French blue + black
Smalt Blue .	Smalt, or French blue + violet ultramarine.
Ultramarine Blue	(Genuine ultramarine, or Italian ultramarine).
Antwerp Blue .	(Antwerp blue)
Campanula Blue	Smalt blue + white
Cobalt Blue . .	Cobalt blue.
China Blue . . .	Antwerp blue + black + white
Flax-flower Blue . .	French blue + white
Azure Blue .	Cobalt blue + white.
———— .	Antwerp blue + white.
Pearl Blue . .	French blue + white.
Sevres Blue . .	Antwerp blue + cobalt blue + white
Glaucous Blue .	Antwerp blue + black + white.
Turquoise Blue	Antwerp blue + cobalt blue + emerald green + white.
Cerulean Blue .	Antwerp blue + cobalt blue + white
Verditer Blue	Antwerp blue + black + light cadmium + white.
Nile Blue . .	Antwerp blue + emerald green + white.

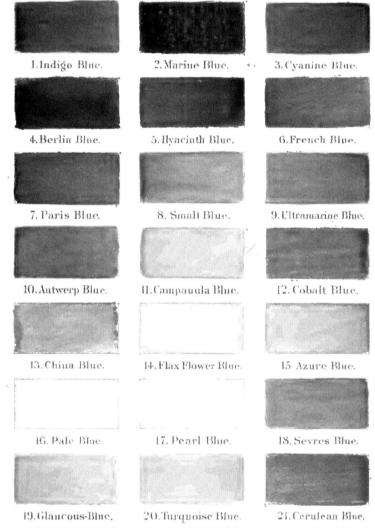

PLATE X.

                          Composed of

Bottle Green . . . Cadmium orange + Antwerp blue.
Myrtle Green . . . Antwerp blue + cadmium orange.
Terre-verte Green . (Terre verte)
Grass Green . . . Cadmium yellow + cadmium orange + Antwerp blue.
Sea Green . . . Viridian + Antwerp blue.
Malachite Green . . Ultramarine blue + pale cadmium + white
Parrot Green . . Antwerp blue + pale cadmium + raw sienna.
Viridian Green . . (Viridian).
Pea Green . Ultramarine blue + pale cadmium + raw sienna + white
Bice Green . Ultramarine blue + pale cadmium + raw sienna + white.
Verdigris Green . . Viridian + pale cadmium + white.
Chromium Green . Ultramarine blue + pale cadmium + black + white
Paris Green . Emerald green + pale cadmium + white.
Beryl Green Viridian + emerald green + Antwerp blue + white.
Sage Green . Ultramarine blue + pale cadmium + black + white.
Emerald Green . (Emerald green).
Glaucous Green . . Viridian + white
Olive-Green . . . Raw sienna + Antwerp blue
French Green Italian ultramarine + pale cadmium.
Apple Green . . Antwerp blue + pale cadmium + white.
Oil Green . . Antwerp blue + pale cadmium + raw sienna.

NOTE.

The **pileum** includes the *forehead*, *crown*, and *occiput*.

The term *cheeks* is more or less indefinite in meaning, but in its widest sense may be said to include the *auriculars, suborbital region,* and *malar region*. It is, however, sometimes restricted to one or the other of these divisions.

The *supraloral region* (15), *superciliary region* (14), and *supra-auricular region* (13) together, or when continuous with one another (as in the figure), constitute a **superciliary stripe.**

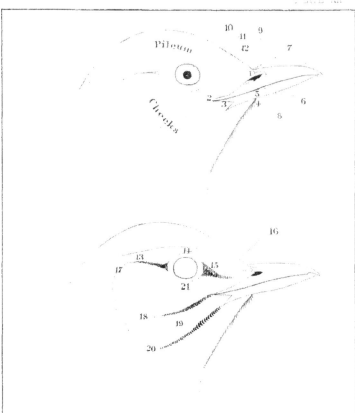

1. Frontal apex.
2. Rictus.
3. Malar apex.
4. Mental apex.
5. Ramus of Mandible.
6. Gonys.
7. Culmen.
8. Nostril.
9. Nasal operculum.
10. Pupil.
11. Iris.
12. Orbital ring.
13. Supra-auricular region.
14. Supra-orbital region.
15. Supra-loral region.
16. Loral streak.
17. Postocular streak.
18. Rictal streak.
19. Malar stripe.
20. Sub-malar streak.
21. Sub-orbital region.

NOTE.

The following divisions include two or more of those distinguished on the figure: (1) The **Upper Parts** comprise the entire upper surface, from *forehead* to *tail*, inclusive, and include also the outer or upper surface of the wings. (2) The **Lower Parts** comprise the entire under surface, from *chin* to *lower tail-coverts*, inclusive, but do not necessarily include the under surface of the wing. The boundary line between the upper and lower parts, on the sides of the head and neck, is indefinite, or variable, and is usually indicated in each particular case in the description of a bird. (3) The **Pileum** includes the *forehead, crown*, and *occiput*. (4) The **Fore-neck** includes the *chin, throat*, and *jugulum*. (5) The **Sides**, in the comprehensive sense of the term, include the *flanks* as well as the *sides* proper. (6) The **Mantle** includes the *back, scapulars*, and outer or upper surface of the wings.

The **Crissum** is properly that portion between the lower tail-coverts and anal region which in the figure is concealed by the primaries. When the *lower tail-coverts* and *crissum* are different in color, they are then distinguished; but when they are concolored, they are usually considered synonymous, the term *crissum* being used for the tract itself, and that of *lower tail-coverts* for the individual feathers.

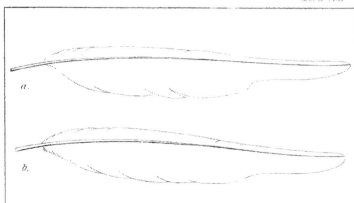

a. An emarginated primary quill.

b. A sinuated primary quill.

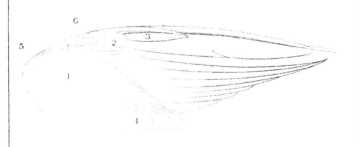

1. Under wing-coverts, or "lining of wing."
2. Under primary-coverts.
3. Spurious primary.
4. Axillars.
5. "Bend of wing" or carpal joint.
6. Carpo-metacarpal joint.

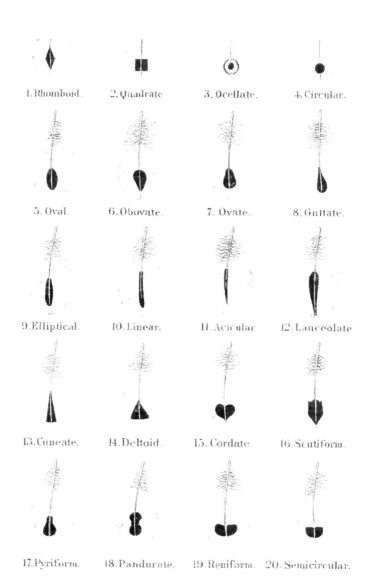

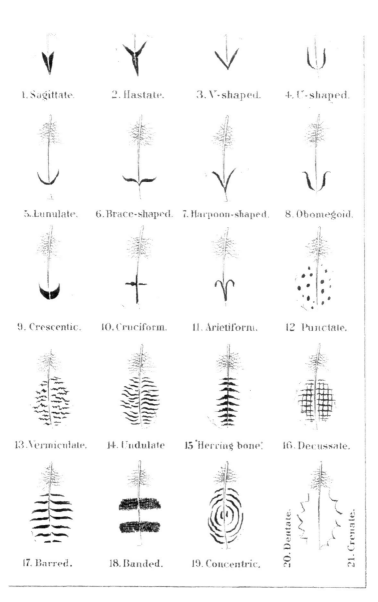

NOTE.

The diagrams shown on this plate, in illustration of the various typical variations in the outline or contour of birds' eggs, are taken, with considerable modification, from a work on Swedish oology, having the following title —

Skandinavisk Oologi. | Utbredning, Bo och Ägg af Sveriges och Norges foglar | jemte | Ornithologisk Exkursions-Fauna | af | Carl Agardh Westerland. | —— | Stockholm. | Albert Bonniers Forlag |

(Published in 1867, 8vo, pll 4, pp. 1-250, i-vi, 1 pl.)

NOTE

Pied du Roi. This standard is used in the works of BONAPARTE, FINSCH, HARTLAUB, SCHLEGEL, and TEMMINCK, and also in those of most of the older French authors

English Duodecimal Used by AUDUBON, MACGILLIVRAY, PALLAS, and all earlier American, English, and Russian authors.

English Decimal Used by BAIRD, and most recent American, English, and Russian authors who have not adopted the metric system.

Metric System. Most modern authors, except those mentioned above.

NAUMANN uses "Leipziger oder gewohnliches Werkmass" The Leipzig foot = 0.9275 English foot, 0 2827 metre, or 0 8703 French foot

The foot used by S Nilsson and other Swedish ornithologists is 0 9742 English foot, 0 9141 French foot, or 0 2969 metre.

The Rhineland foot (which is the same as the Prussian, Rotterdam, and Danish foot) is frequently used by the earlier German authors It corresponds to 1.0298 English feet, 0.9663 French foot, and 0.3139 metre

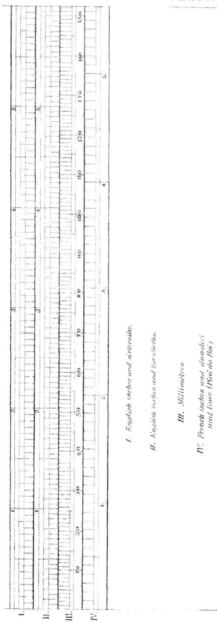

Milton Keynes UK
Ingram Content Group UK Ltd.
UKHW021838031123
431868UK00004B/57